Other Titles in the Science of Pop Culture Series

The Science of Michael Crichton (March 2008)

*An Unauthorized Exploration into the Real Science
Behind Frank Herbert's Fictional Universe*

THE SCIENCE OF DUNE

EDITED BY KEVIN R. GRAZIER, PH.D.

BENBELLA

BENBELLA BOOKS, INC.

Dallas, Texas

"Melange" © 2008 by Carol Hart, Ph.D.
"My Second Sight" © 2008 by Sergio Pistoi, Ph.D.
"The Biology of the Sandworm" © 2008 by Sibylle Hechtel, Ph.D.
"The Dunes of Dune" © 2008 by Ralph D. Lorenz, Ph.D.
"From Silver Fox to Kwisatz Haderach" © 2008 by Carol Hart, Ph.D.
"Evolution by Any Means on Dune" © 2008 by Sandy Field, Ph.D.
"The Anthropology of *Dune*" © 2008 by Sharlotte Neely, Ph.D.
"The Real Stars of Dune" © 2008 by Kevin R. Grazier, Ph.D.
"Prescience and Prophecy" © 2008 by Csilla Csori
"Stillsuit" © 2008 by John C. Smith
"The Black Hole of Pain" © 2008 by Carol Hart, Ph.D.
"Navigators and the Spacing Guild" © 2008 by John C. Smith
"Memory (and the Tleilaxu) Makes the Man" © 2008 by Csilla Csori
"Cosmic Origami" © 2008 by Kevin R. Grazier, Ph.D.
"Suspensor of Disbelief" © 2008 by Ges Seger with Kevin R. Grazier, Ph.D.
"The Shade of Uliet" © 2008 by David M. Lawrence
Additional Materials © 2008 by Kevin R. Grazier, Ph.D.

BenBella Books, Inc.
6440 N. Central Expressway, Suite 503
Dallas, TX 75206
www.benbellabooks.com
Send feedback to feedback@benbellabooks.com

Printed in the United States of America
10 9 8 7 6 5 4 3 2 1

Library of Congress Cataloging-in-Publication Data

The science of Dune : unauthorized exploration into the real science behind Frank Herbert's fictional universe / edited by Kevin R. Grazier, Ph.D.
   p. cm.
  ISBN 1-933771-28-3
  1. Herbert, Frank. Dune. 2. Science fiction, American--History and criticism. 3. Dune (Imaginary place) I. Grazier, Kevin Robert, 1961-
  PS3558.E63Z87 2007
  813'.54—dc22

2007036294

Proofreading by Stacia Seaman and Jennifer Canzoneri
Cover design by Laura Watkins
Text design and composition by John Reinhardt Book Design

Distributed by Independent Publishers Group
To order call (800) 888-4741
www.ipgbook.com

For special sales contact Robyn White at robyn@benbellabooks.com

CONTENTS

INTRODUCTION

Kevin R. Grazier, Ph.D.

SINCE WE LIVE IN what is called "The Information Age" it didn't surprise me when, long before I started writing this introduction, I saw advanced PR for this book online. I was also not surprised when a friend pointed out that I could go to Amazon.com, see the book cover and, of course, pre-order a copy—well before I'd even finished my first essay. I wasn't particularly surprised when, a week later, I found a talkback board online, rife with speculation about what this book might contain, how good it might be, and whether or not it would be "worth it." What did jump out at me from some inky black abyss was a brief entry on one chat board that read simply, "Is there any science in *Dune*?"

My immediate reaction was a predictable, "What kind of question is that? Of course there's science in *Dune*, you moron!" Hang on. When I re-explored this idea with more consideration, and less knee-jerk, my response was far less critical. Now, certainly, if there was no science in Frank Herbert's Dune novels, the book in your hands would rival *Everything Men Know about Women* as the world's shortest. On the other hand Frank Herbert, either by design or accident, mostly gave hints about the technical underpinnings of the Dune Universe (I prefer the term Duniverse) employing, instead, a technique I'll call *science by allusion*. Herbert avoided the pitfall into which many science fiction writers wander: by omitting much in the way of technical details, the work never becomes dated, at least in this respect.

This is not unimportant. Technology is an increasingly important part of all our lives, and we are routinely exposed to technology today undreamt of in 1965 when *Dune* was published. A detailed description of an "average" person's day in 2008 may include references to the Internet, personal computers, recycling, cell phones, wi-fi, ATMs,

Segways, satellite TV, and other technologies that would be identified as science fiction forty years ago. We find a prime example when we consider the first season of *Star Trek* which, coincidentally, premiered in 1965. Their twenty-third-century communicators were considered science fiction then, yet the technology in our flip-open cell phones has clearly outpaced their *Star Trek* counterparts—for we never got the impression that playing Tetris on his communicator was an option for Mr. Spock, nor did Captain Kirk ever get a text message during a meeting. When a writer makes a technical gaffe, the increasingly technically literate reader of today is taken out of the novel, is no longer seeing the depths of the writer's universe through the eyes of one of its characters, and reverts instead to a person in the twenty-first century holding a book saying, "Hang on a minute!"

Like any description of everyday tech from 2008 would seem as science fiction in 1965, Herbert's description of the Duniverse necessarily included allusions to their society's science and technology. Herbert created a drug called melange, or spice, to endow both Guild Navigators and the Bene Gesserit with their unique abilities, and to endow the affluent with extended life. He created giant sandworms to, in turn, create the spice. He created the nebulously defined Holtzmann Effect as the basis for suspensor, shield, glowglobe, and even FTL technology. In fact, in one of the few cases where he does provide limited details (the topic of stillsuits) he actually gets into a bit of trouble, as we will explore in these pages.

While all these topics are described in *Dune*, the work isn't *about* any of these things. *Dune* explores themes both vast and epic: politics, conquest, religion, ecology, and one individual's personal journey from man to superman. By, for the most part, merely alluding to the science and technology—and glossing over details—Herbert avoids technical gaffes, does nothing to take his reader out of the work, and it remains as accessible, enjoyable, and as pertinent today as it did when it was first published. It can be argued successfully, in fact, that *Dune* is more pertinent today than it was in 1965 on some levels. Many areas of science to which Dune alludes (examples are: genetic manipulation, exoplanets, extremophile life forms, comparative planetology) were the realm of science fiction—or at best speculative science—in 1965, but are very hot topics today.

For the technophiles among us, who thrive on explaining how things work, Herbert left a whole sandbox full of toys to enjoy. The essays you hold run an entire gamut. Some ("The Real Stars of Dune," "The Dunes of Dune," "The Shade of Uliet") explore and elaborate upon the environments within the Duniverse. Other essays ("Suspensor of Disbelief," "My Second Sight," "From Silver Fox to Kwisatz Haderach") provide technical underpinnings of the Duniverse that Herbert never supplied, yet on the other hand, some (the previously mentioned "Stillsuit" and "Cosmic Origami") question whether some of the Technology of the Imperium will ever come to fruition. Still others ("Memory [and the Tleilaxu] Makes the Man," "The Biology of the Sandworm") can be stamped "TBD." I hope, dear reader, that you enjoy reading the following series of essays as much as we, the authors, have enjoyed playing in Frank Herbert's sandbox, ensuring that, yes, there is now definitely science in *Dune*.

MELANGE

Carol Hart, Ph.D.

With respect to the addictive spice melange, *"Just say 'No'" is apparently not a phrase known within the Imperium. Why should it be, when spice enables faster-than-light travel, allows the Bene Gesserit their insight, and extends life for all? At what price, though? Carol Hart, Ph.D., explores some of the consequences of mind-expanding substances as well as their unintended side-effects.*

LONG AGO AND FAR AWAY, I went to an experimental college in Florida, where the experimenting was not just with teaching methods and grading systems but with mind-altering drugs. Alcohol use was frowned upon. So bourgeois! So '50s! But just about everyone smoked pot, and perhaps a quarter of the student body had dropped acid or tried other psychoactive drugs. My roommate Margot was on a Carlos Castaneda-inspired quest for enlightenment via her drug of choice, LSD. One afternoon she walked into our room, her green eyes still glassy and staring, and told me about her latest mind-bending discovery. She had walked for miles upon miles along the bay while she was tripping and had talked to the mosquitoes. She asked them not to bite and they agreed to spare her. She held up her bare, unblemished arms as proof—see, no bites!

Walking along that bay with its bug-infested backwaters and coming back without a half-dozen bites would be a near miracle. From a rationalist perspective, there were some obvious explanations that occurred to me at the time. Perhaps Margot had been out for so many hours that the swelling had gone down, and she simply hadn't felt or noticed the bites while she was tripping. Or perhaps the drug had affected her skin chemistry and the composition of her sweat so that the mosquitoes were not attracted to her. I kept my skepticism to myself, made some polite "Wow! Cool!" remarks, and went back to my studying.

After rereading *Dune* and thinking about the properties of melange, another perspective on Margot and her friendly mosquitoes suggests itself. The Bene Gesserit or the Guild Navigators might say, "That's it? You took a powerful consciousness-expanding, space- and time-warping drug, and all you used it for was preventing bug bites?"

Their scornful superiority would be a bit unfair: merely reading Castaneda could not have given Margot the knowledge and training needed to send her awareness soaring through time and space as they do. Aside from a vague desire for out-of-body experiences (and a specific wish to be itch-free when she came back to that body), Margot had no goals for her LSD adventures.

Albert Hofmann, the discoverer of LSD, always lamented the casual and recreational use of hallucinogenic agents. He had hoped that experimenting with LSD would take place as formal research involving neuropsychiatrists, theologians, and artists engaged in serious exploration of perception, religious experience, and the unconscious self. "Deliberate provocation of mystical experience, particularly by LSD and related hallucinogens, in contrast to spontaneous visionary experiences, entails dangers that must not be underestimated," he wrote in *LSD: My Problem Child*. "Special internal and external advance preparations are required; with them, an LSD experiment can become a meaningful experience" (32). Both the melange adepts of *Dune* and the shamans of Old Earth would likely agree with Hofmann's assessment.

The Roles of Ritual and Expectation: Shamans versus Day-Trippers

On first reading, one of the more SF aspects of melange is the broad spectrum of its users and uses—a true *mélange* of effects. For the Guild Navigators, melange allowed them to quest through time to find the paths on which to guide the Heighliners. For the Bene Gesserit, the agony induced by spice essence initiated a Reverend Mother's communion with the Other Memory of all her forebears. For the Fremen, the spice was a daily food and (in the form of the Water of Life) the aphrodisiac that powered their sietch orgies. For the wealthier classes of the Empire, small daily doses vastly extended life expectancy.

Here on Old Earth, we have never discovered a substance to extend life expectancy (not for want of trying). But we do possess a number of natural psychoactive substances that share features of melange. Fantastic as the effects of melange may seem, there are some equally impressive parallels among our own mind-altering agents: marijuana, opium, coca leaves (cocaine), ayahuasca (dimethyltryptamine, DMT), magic mushrooms (psilocybin and psilocin), and peyote (mescaline). Like melange, these hallucinogens have traditionally been used to transcend time and place, to understand the distant ancestral past, and to peer into the bifurcating paths of destiny. Like melange, their functions were medicinal, ritual, and political.

In the Amazonian rainforest, ayahuasca (Vine of the Soul, also known as *yajé*) is used by shamans to communicate with the spirit world, to diagnose the supernatural causes of illness, and to give them the prescience needed to solve problems confronting the tribe, such as perceiving the battle plans of an enemy. Ayahuasca is also taken ritually by the men of the community, who chant and sing of the visions that unfold under the power of the sacred drink. Because of the prescient qualities of the visions as experienced by the Amazonian peoples, one of the principal psychoactive compounds isolated from ayahuasca was originally named *telepathine*, until it was found to be identical with the already named *harmine*, found in another hallucinogenic plant, Syrian rue.

The traditional process of preparing the Vine of the Soul has an ingenuity comparable to the Freman method of distilling the Water of Life by drowning a stunted sandworm or "little maker" to produce a liquid that then must be detoxified by a Reverend Mother. The power of the Amazonian brew depends upon the synergistic combination of plants containing *tryptamine derivatives* (primarily DMT, dimethyltryptamine, a powerful hallucinogen) with those containing psychoactive *beta-carboline alkaloids* (harmine, harmaline). When taken orally, the tryptamines would normally be inactivated by monoamine oxidase in the liver and gastrointestinal tract. However, the beta-carboline alkaloids are natural analogs to a pre-Prozac class of antidepressants known as the *monoamine oxidase inhibitors*. The presence of the beta-carboline compounds (which have milder psychoactive properties) allows the more potent tryptamines to escape digestion and pass through the blood-brain barrier. For all the claims made about molecularly targeted drugs, pharmaceutical chemists have rarely designed a drug as sophisticated as the ayahuasca of the Amazonian peoples.

The mystique of ayahuasca has attracted tourists ready to pay a high premium for an authentic shamanistic experience, as well as hothouse efforts in temperate North America to grow the vines. Outside of its ritual uses, when taken by researchers or novelty seekers ayahuasca has unpredictable effects that can be overwhelmingly negative. Yet, to the indigenous peoples of the rainforest, it enables visions of the gods, the creation of the world, and the ancestral past, as well as telepathic contact with others. Their expectation of a revelation gives them the strength to withstand the vine's equivalent of the "spice agony," the jarring physical symptoms of pain, nausea, vertigo, and prostration that precede the liberating visions. Outside of the cultural context that gives them meaning, the visions produced by ayahuasca can be disturbing and disorienting in the extreme if, when everyday reality is stripped away, the consciousness is unprepared to interpret what takes its place.

As experienced by ethnobotanist Wade Davis, ayahuasca has a power and terror approaching melange: "Soon the world as I knew it no longer existed. Reality was not distorted; it was dissolved as the terror of another dimension swept over the senses . . . the terror grew stronger, as did my sense of hopeless fragility. Death hovered all around . . . my thoughts themselves turned into visions, not of things or places but of an entire dimension that in the moment seemed not only real, but absolute. This was the actual world, and what I had known until then was a crude and opaque facsimile" (*Shadows in the Sun* 157).

Ethnobotanists stress the role of cultural and individual expectations in conditioning the hallucinogenic experience. From his own experience and that of others, Wade Davis wisely commented that "the pharmacologically active components do not produce uniform effects. On the contrary, any psychoactive drug has within it a completely ambivalent potential for good or evil, order or chaos. Pharmacologically, it induces a certain condition, but that condition is mere raw material to be worked by particular cultural or psychological forces" (*Shadows in the Sun* 166). Similarly, Michael Balick and Paul Alan Cox wrote that "there is considerable reason to believe that the use of hallucinogenic plants outside of their traditional religious contexts can produce sorrow rather than transcendence, confusion rather than enlightenment" (*Plants, People, and Culture* 159).

That religious context can both direct and heighten the experience. Many of our common legal psychoactive plants were once reserved for ceremonial use, including tobacco (think of the Native American peace pipe) and the currently popular stimulant kava. "The real power of kava," said Balick and Cox, "comes from the cultural context in which it is drunk. Consumed in one's home, kava has an effect that is scarcely noticeable. But drinking kava under a thatched roof ten meters high in the presence of the assembled chiefs of the entire district, all of whom scrupulously follow the ancient forms of rhetoric, is a truly memorable experience" (*Plants, People, and Culture* 162).

While melange's effects are clearly more in line with the most powerful awareness-enhancing agents we possess, its power also is greatly influenced by mindset and setting. Billions of the more affluent citizens of the Empire ingested it daily, apparently without experiencing its enhanced awareness-spectrum effects. Possibly it was to them as coffee, tea, kava, and tobacco are to us—a mildly addictive psychostimulant when habitually consumed without ritual or training to focus its effects. As with ayahuasca, peyote, and LSD, "the pharmacologically active components do not produce uniform effects" at higher levels of exposure. Piter, Baron Harkonnen's Mentat, had the dark blue-tinted eyes of a high-dose user (the Baron grumbled that Piter ate spice like candy), yet his melange use did not seem to grant him prescience. He was unable to foresee the escape of Jessica and Paul, or Yueh's double act of treachery that cost Piter his life. A sadistic sociopath, Piter is described as a "twisted" Mentat, with no explanation other than the brief *Dune* glossary entry that twisted Mentats were a specialty product of the Tleilaxu. It is just possible that Piter's dependence on melange as a recreational drug (he referred to it as one of his expensive "pleasures" [*Dune* 17]) was the vehicle for corrupting and degrading him.

In sharp contrast, melange endowed Paul Atreides, who also possessed Mentat abilities, with a vision far beyond that achieved by the Guild or the Reverend Mothers. Paul did not attribute his immense prescience to genetics, to the Bene Gesserit breeding program, but to the training given to him and the timing of his exposure to the spice. "He could look to his own past and see the start of it—the training, the sharpening of talents, the refined pressures of sophisticated disciplines, even exposure to the O.C. Bible at a critical moment...and, lastly the heavy intake of spice" (*Dune* 194–195). However great its potency for the prepared mind, the overall effects of melange have several distinct parallels to our own terrestrial hallucinogens:

- *Telltale changes to the eye.* Hallucinogens such as LSD and ayahuasca produce extreme dilation of the pupil; from their daily large doses of spice, the Fremen and the Guild had a deep dark blue coloring to the sclera.
- *Suspension of time.* Seconds become hours, or time stops altogether. Paul was unconscious for three weeks when he made his experiment with changing the Water of Life, but believed it had only been a few minutes.
- *Ecstatic (or sometimes frightening) sense of communion with others.* Depending on the cultural context, this mystic communion might be with ancestral spirits (the Other Memory of the Bene Gesserit), or gods, or with the universe itself.
- *Out-of-body sensations.* Jessica "felt that she was a conscious mote, smaller than any subatomic particle" (*Dune* 354).
- *Loss of self and merger into a oneness.* Jessica experienced something she called a "psychokinesthetic extension" of self (*Dune* 354).
- *Euphoria.* A major component of the experience with terrestrial hallucinogens, euphoria was part of Jessica's experience of the Water of Life. "The muzziness of the drug was overpowering her senses. How warm it was and soothing. How beneficent of these Fremen to bring her into the fold of such companionship" (*Dune* 359).
- *Death-rebirth experience.* This is Hofmann's characterization of the progression from physical agony or terror and to euphoria that is characteristic of LSD and related hallucinogens. The spice agony that a Bene Gesserit Reverend Mother had to undergo seemed to follow a similar course from pain (the spice essence could be fatal) to awakening to a shared communion with her predecessors.
- *Visions/hallucinations.* Phantasmagoric shifts of color and shape are typical of LSD, but the experience may bring a sense of traveling through space or time to visualize ancient cities or strange civilizations. Sometimes a voice comments upon or interprets the hallu-

cinogenic experience. It seems that the Other Memory of the Bene Gesserit was not the product of their discipline so much as it was the gift of the spice. At the moment of his death, planetologist Liet-Kynes, who had the eyes of a melange user, had his own experiences with Other Memory (his hallucination of his father).

- *Prescience and life-changing realizations.* For the Amazonian shamans, ayahuasca allowed the soul to leave the body to search out the explanation for illness in the individual or problems threatening the community and to decide the course of action. When Paul tasted the unchanged Water of Life, he saw the ships of the Guild filling the space above Arrakis and understood why they were there—that they, too, were addicted to the spice—and knew what he needed to do to reclaim Arrakis for House Atreides.

The Biochemistry of Transcendence

The experience of LSD both inspired and sabotaged serious academic research into the mechanisms of hallucinogenic agents such as ayahuasca. Seventy years ago, Albert Hofmann, a Sandoz research chemist, was engaged in a project to synthesize and screen derivatives of ergot, a fungus found on rye, for medicinal properties. The twenty-fifth compound in his series, lysergic acid diethylamide (LSD) failed to demonstrate any unique properties in the initial laboratory and animal studies. The ergot screening project had already identified several promising drug candidates, including the anti-migraine drug ergotamine, still in use today. LSD seemed less active than a related compound, ergonovine, seen as having therapeutic potential for preventing postpartum hemorrhaging. So it was dropped from further clinical development by Sandoz.

Five years later, in 1943, acting on "a peculiar presentiment," Hofmann decided to synthesize and screen LSD once more (*LSD: My Problem Child* 12). Later that day, he noticed some odd but distinctly pleasant visual effects and suspected an accidental exposure to the novel compound as the cause. So he took what he considered to be a miniscule dose of the agent, just 0.25 mg (which was five to ten times the doses taken by my roommate Margot). After an unsteady bicycle ride home with his laboratory assistant to guide him, he was plunged into terrifying hallucinations in which he feared for his sanity and his life. "A demon had invaded me, had taken possession of my body, mind, and soul. I jumped up and screamed, trying to free myself from him, but then sank down again and lay helpless on the sofa" (*LSD: My Problem Child* 15). The reassurance of a doctor that his vital signs were normal, with extremely dilated pupils as the only abnormality, helped him through the crisis, and then the hallucinations became more pleasant.

> Kaleidoscopic, fantastic images surged in on me, alternating, variegated, opening and then closing themselves in circles and spirals, exploding in colored fountains, rearranging and hybridizing themselves in constant flux. It was particularly remarkable how every acoustic perception, such as the sound of a door handle or a passing automobile, became transformed into optical perceptions. Every sound generated a vividly changing image, with its own consistent form and color (*LSD: My Problem Child* 16).

Like a good scientist, Hofmann made a full report of his experiences. His colleagues were skeptical that such a minute quantity of a compound could produce such extreme effects, until they tried it themselves. Under the trade name of Delysid, Sandoz made LSD available for researchers to explore two potential, rather contradictory indications: first, as a psychiatric treatment "to elicit release of repressed material and provide mental relaxation, particularly in anxiety states and obsessional neuroses" (*LSD: My Problem Child* 3) and second, as a model for studying experimentally induced psychosis. Hofmann himself hoped the drug would be a modern ayahuasca, a vehicle for enlightenment and cure used only under the guidance of disciplined professionals, the scientific equivalent of shamans and Reverend Mothers.

Just a few years later, *serotonin* (5-HT, 5-hydroxytriptamine) was discovered, first in blood serum, then in various tissue samples, and finally in the brain. Once the chemical structures had been elucidated, it was quickly noted that two of the four rings in LSD's chemical structure are identical to the ring structure in serotonin, and the side chain attached to serotonin's ring structure is identical to another part of the LSD molecule. The similarities were so striking that suspicion was immediately raised that LSD might be exerting its profound hallucinogenic effects by mimicking, augmenting, or displacing serotonin in the central nervous system.

The discovery of LSD spurred interest in the chemistry of plant hallucinogens being gathered by botanists and anthropologists in the Americas. It was Albert Hofmann who successfully isolated the psychoactive components of the Mexican magic mushroom (*Psilocybe mexicana*) in 1958, fifteen years after his first experiments with LSD. He found that psilocybin and psilocybin are also structurally very similar to both serotonin and LSD. When he tested them on himself, he reported the effects to be so similar that LSD is virtually an extra-strength version of the fungus. Many of the plant hallucinogens, including DMT, the major psychoactive compound of ayahuasca, are tryptamines, a class of compounds related to the amino acid *tryptophan* and its product, the neurotransmitter serotonin.

Knowing that these psychedelic compounds closely resembled serotonin was only a first step to understanding how these drugs act on the brain. Serotonin has bafflingly complex roles in different parts of the brain and body. As a neurotransmitter, serotonin modulates mood, pain awareness, impulse control, aggression, vomiting, and appetite. Outside the brain, serotonin regulates contraction and relaxation of the smooth muscle lining the blood vessels and the intestines, and triggers blood platelets to stick together to make a clot. Abnormalities in the serotonin system underlie depression, anxiety, eating disorders (bulimia, bingeing, anorexia nervosa), alcoholism and other addictions, migraine, fibromyalgia, and irritable bowel syndrome.

Like other hormones and neurotransmitters, serotonin is released by one cell to trigger a specific response in a target cell by interacting with a molecular receptor on the cell surface. This interaction triggers a cascade of internal signals that result in some specific change in the cell's activity. Serotonin itself is a very simple molecule that is found throughout the animal kingdom, from garden slugs to primates, and also in many plants. The complexity of the serotonin signaling system results from the multiplicity of receptor types on target cells. There are fifteen known receptor subtypes grouped into seven families, from $5\text{-}HT_{1A}$ to $5\text{-}HT_7$. Clearly, any drug that hits receptors for a neurotransmitter with so many duties—from pain perception and mood regulation to gut peristalsis—can have multiple effects. Action on the serotonin system would readily explain why LSD can produce its roller coaster of emotional states, from terror and despair to exaltation within a few hours. Ecstasy (MDMA) produces euphoria without psychedelia (or only mild visual effects) by flooding the brain with serotonin. But it is by no means obvious why a serotonin mimic would produce hallucinations as one of its effects.

There are other serotonin-like substances in the brain and body that are even closer structurally to the hallucinogens: tyramine, tryptamine, octopamine, beta-phenylethylamine (which acts as an amphetamine), and several psychoactive forms of tryptamine, including 5-methoxy-tryptamine and DMT, the major psychoactive component of ayahuasca. Because these compounds are found only in trace amounts in brain tissue samples, most researchers have classified them as metabolic by-products of the synthesis of the major neurotransmitters (serotonin, *dopamine*, and *norepinephrine*) and dismissed them as unimportant. Recent studies have identified specific receptors for these so-called *trace amines*, demonstrating that these chemicals likely have real, if still unknown, roles in how information is processed in the brain. Some still sketchy findings suggest that they may trigger or enhance the release of serotonin and the related neurotransmitters dopamine and norepinephrine. Given its relative abundance (and the relative ease of studying it), serotonin is still the lead candidate as the primary target of the hallucinogens.

The heavy-duty science of unraveling the functions of the fifteen subtypes of serotonin receptors has been greatly eased by the development of genetically engineered mice—known as knockout mice—that lack the gene for the protein of interest. We know from behavioral studies with these animals that the 5-HT_{1A} knockout mouse is a timid quitter, whereas the 5-HT_{1B} knockout mouse is an aggressive bully and a heavy drinker. The 5-HT_{5A} knockout mouse shows reduced sensitivity to LSD. The functioning of this specific receptor is not yet well understood, but what is known is quite suggestive. Based on studies of its localization in the brain, the 5-HT_{5A} receptor may have regulatory roles in the brain's internal timekeeping as well as mood and cognitive function. Interestingly, genetic variants of this receptor are implicated in schizophrenia. Receptors in the 5-HT_2 class are also targeted by the hallucinogens, with a clear association between hallucinogenic potency and the level of receptor binding. Based on animal models, it has been proposed that activation of 5-HT_{2A} receptors by hallucinogens triggers a hypersensitivity to sensory information in the *locus ceruleus*, an area of the brainstem that has overall responsibilities for interpreting and responding to environmental threats.

Animal models for studying hallucinogenic effects have some obvious limitations: mice cannot be interviewed about their LSD experiences, so the effects must be inferred by behavioral changes. The widespread recreational use of LSD in the 1960s and its subsequent labeling as a drug of abuse in most countries resulted in a cessation of clinical research. But there are people who hallucinate under perfectly legal and respectable circumstances. In other ages they were revered as saints, shamans, and seers. In ours they tend to keep their experiences to themselves.

Hearing Voices, Seeing Visions

"Think you of the fact that a deaf person cannot hear. Then, what deafness may we not all possess? What senses do we lack that we cannot see and cannot hear another world all around us?"

—Orange Catholic Bible (*Dune* 40)

These were the first words Paul Atreides read in the Orange Catholic Bible given to him by Dr. Yueh on the eve of Paul's departure from his home planet of Caladan. On Arrakis, as he first awakened to his terrifying powers of prescience, he recognized this timely exposure to the O.C. Bible as part of the "clockwork" of experiences and training that brought him to this moment of his transformation (194). Still later, when he piloted the 'thopter into the sandstorm to ride it out on a vortex, he felt himself on the terrifying brink of another revelation. It came to him out of the words of the Bene Gesserit fear litany ("I shall not fear"), and the now-familiar words of the Orange Catholic Bible still rang in his memory: "What senses do we lack that we cannot see or hear another world all around us?"

In "Appendix II: The Religion of Dune," we learned that the O.C. Bible was the ecumenical compilation of the "fourteen sages" representing the major religions of humanity, including the ancient teachings of the Zensunni Wanderers, the Navachristianity of Chusuk, and the Buddislamic Variants. The verse that Paul read might well have been inspired by any of these traditions, for it encapsulates a perception that repeats itself in many forms: that this everyday world, so vividly and concretely real to our senses, is only the dim shadow of some greater reality.

"For now we see through a glass, darkly; but then face to face: now I know in part; but then shall I know even as also I am known."

—Paul, Corinthians 1:13

"If the doors of perception were cleansed everything would appear to man as it is, infinite. For man has closed himself up, till he sees all things thro' narrow chinks of his cavern."

—William Blake, *The Marriage of Heaven and Hell,* 14

"This world that appears to the senses has no true being, but only a ceaseless becoming; it is, and it also is not; and its comprehension is not so much a knowledge as an illusion."

—Arthur Schopenhauer, *The World as Will and Representation,* Vol. I, Appendix: "Criticism of the Kantian Philosophy"

It is a perception that has been affirmed by many experimenters with ayahuasca, mescaline, and LSD, and it primed Paul for his own drug-induced experience. It is not, after all, soft mysticism but hard science: the world our senses perceive is an illusion or, at best, a blinkered squint at a glittering, buzzing, whirling reality beyond our comprehension. We perceive a few spectra of electromagnetic radiation and name them as colors; what we experience as solidity under our fingertips is another manifestation of indivisible energy, the dance of subatomic particles in space.

Philosophical reflections and hallucinogenic experiences aside, doubting our sober everyday realities is the slippery slope of insanity. Both in popular culture and in the psychiatric consulting room, people who hallucinate are presumed to be psychotic. If not for the causative factor of their hallucinogenic drug exposure, both Paul and Jessica would meet the diagnostic criteria for schizophrenia. They show so-cial/occupational dysfunction in their rapid downward spiral after los-ing control of Arrakis to Baron Harkonnen. Paul's perceptions would certainly qualify as bizarre delusions to any right-thinking psychia-trist. The Other Memory of the Fremen Reverend Mothers that Jessica accessed through the Water of Life ritual constitutes the classic symp-tom of schizophrenia: "A voice keeping up a running commentary on the person's behavior or thoughts, or two or more voices conversing with each other" (*Study Guide to DSM-IV* 140).

Someone who sees angels, aliens, or ghosts may be tolerated as a harmless crank. People who hear voices are considered potentially dan-gerous to themselves and others. Auditory hallucinations are the stereo-typical proof of madness. No one who is otherwise tightly wrapped will publicly admit to hearing voices, yet it may be fairly common. In stud-ies, 2 to 4 percent of the general population say that they sometimes hear voices when no one is there, with two-thirds of them not meeting other diagnostic criteria for mental illness. For some individuals who do have such a diagnosis, distress at hearing voices may be their major motive for seeking help. A small number of mental health profession-als are now exploring the therapeutic value of accepting auditory hal-lucinations rather than tranquillizing them into silence. This alternative approach suggests that voices are only a psychiatric issue when the indi-vidual has difficulty coping with them. Marius Romme and colleagues at Maastricht University have reported that the auditory hallucinations ex-perienced by ordinary people are not substantially different from those of schizophrenic patients. The major difference was that ordinary peo-ple tended to perceive the voices as positive and felt in control of the experience. It did not matter how they explained the voice—whether they recognized it as an unconscious projection or believed it to be their guardian angel or perhaps the invisible friend who had been with them from childhood. So long as they perceived the voice as providing guid-ance rather than overwhelming them with inescapable criticism or de-mands, they were unlikely to have other psychiatric symptoms.

In *Dune* and its sequels, those who heard the voices of Other Memo-ry had to control them in order to benefit from their guidance. Other-wise, they became a cacophony of competing personalities that could overwhelm the individual's sense of self and paralyze decision-making.

In *Children of Dune*, Jessica's daughter Alia, who was exposed to Other Memory while still in the womb, ultimately failed to control her voices: "Again the other lives within her lifted their clamor. The tide once more threatened to engulf her" (*Dune* 61). Alia succumbed to the manipulative and malevolent presence of the evil Baron Harkonnen and plunged into madness.

Transcendence, Politics, and Dependence

We live in a culture where hallucinogenic drugs are illegal, feared for their potential powers of enslaving the body while liberating the mind. On Old Earth and in the Empire of *Dune*, the attractions of psychoactive drugs have also been effective means of domination. In pre-Columbian America, the coca leaf was, somewhat like melange, largely reserved for the noble and priestly classes of the ancient Incas. In fact, the ruling classes retained their power in part by their monopoly on the coca leaf. After a period of civil wars and insurrection, the monopoly weakened and coca cultivation spread. Ultimately, the power of the coca leaf was turned against the Native Americans when their new masters, the Spaniards, realized that its consumption would enable their slave workers to overcome hunger, thirst, and altitude sickness, thereby improving their productivity.

It would be missing one of Herbert's major concerns not to see that melange is a means for controlling both individuals and whole societies. At the opening of *Dune*, melange is simply the geriatric spice, available only to the very wealthy, with no one as yet suspecting its importance to the Guild or its addictive potential. In *God Emperor of Dune*, Leto II both hoarded spice and made it more widely available, thereby increasing his tyrannical hold over the far-flung population of his empire. Baron Harkonnen controlled his Mentat Piter by his spice addiction. When Piter was killed, the Baron quickly and coolly replaced him as captain of the guard with Nefud, also an addict, but to the music-drug combination semuta. "A useful item of information, that," the Baron thought to himself in making his decision to promote Nefud (*Dune* 184).

In the brief preface to *Heretics of Dune*, Herbert listed the interlocking themes he had in mind when he sat down to write *Dune* after six years of preliminary research. As one of those themes: "It was to have an awareness drug in it and tell what could happen through dependence on such a substance" (*Heretics of Dune* v). The dependence was physical and psychological, both individual and societal. According to the analysis provided by the condemned historian Bronso of IX at the opening of *Dune Messiah*, "As with all things sacred, it [melange] gives with one hand and takes with the other. It extends life and allows the adept to foresee his future, but it ties him to a cruel addiction and marks his eyes as yours [the priest-inquisitor's] are marked: total blue without any white. Your eyes, your organs of sight, become one thing without contrast, a single view" (*Dune Messiah* 2). Although these remarks were angrily condemned as heresy, they mirrored Paul Maud'Dib's own assessment.

At the moment that Paul achieved his prescience under the pervasive power of the planet's spice, he realized it was, as he told his mother, a subtle poison, condemning them both to life on Arrakis. Except when he first shared the Water of Life with Chani, Paul never experienced spice intoxication as positive. Shortly before he took his first ride on a sandworm, under the lingering spell of a meal heavy in spice, Paul had difficulty sorting past, present, and future events into their proper order—in effect, asking himself, "When am I?" much as the more familiar sort of dazed addict might ask, "Where am I?"

The prescience itself was not a joyful or triumphant experience for Paul or others of the Atreides. The words at the end of Book One of *Dune* in which Paul announced to his mother how the spice has given him prescience are loaded with bitterness and anger: "The spice changes anyone who gets this much of it, but thanks to *you*, I could bring the change to consciousness. I don't get to leave it in the unconscious where the disturbance can be blanked out. I can *see* it" (196). Later, as the Freman shared the Water of Life, Paul realized that the people of the desert also had the capacity for the visions granted by the spice, "*But they suppress it because it terrifies*" (*Dune* 361). His prescience was a prison for Paul, robbing him of all freedom to choose. Toward the end of *Dune Messiah*, he cried out to the Duncan ghola: "I'm dying of prescience.... People call it a power, a gift. It's an affliction! It won't let me leave my life where I found it" (*Dune Messiah* 256).

In his journals, Paul's half-worm, half-human son Leto II made the point more forcefully (as he does everything): "What is the most profound difference between us, between you [the reader] and me? You already know it. It's these ancestral memories. Mine come at me in the full glare of awareness. Yours work from your blind side. Some call it instinct or fate" (*God Emperor of Dune* 103). Before the Baron took over her consciousness in *Children of Dune*, Alia wished for the blindness to ancestral memory that most others had, "living only the hypnoidal half-life into which birth-shock precipitated most humans" (*Children of Dune* 14). To paraphrase ethnobotanists Balick and Cox, it seems that transcendence can bring sorrow, enlightenment ultimately can be confusion.

What Dreams May Come?

My skepticism regarding Margot's LSD-forged friendship with the mosquitoes was somewhat forced. Actually, I admired Margot for her calm, spacey poise and her otherworldly sophistication. I wanted to be more like her and less like me: a gawky rube from western Arkansas. I bought a tab of acid from her and stuck it in among my toiletries, where I looked at it every morning (it was so small to possess such power!) for a couple of weeks, wondering whether I would in fact have the courage to take it.

Growing up among mountains, I had never seen the ocean or any body of water larger than a minor lake. When I first walked across campus to look out on the bay, my spirit left my body, soared across the water to the slender green keys of the distant shore, then came back to me. It happened in one brief gasp of an instant—I barely thought, "I am out of my body!"—and it was already over, never to happen again. I contemplated that little tab of acid every morning with a powerful longing to get that moment back, to hold on to it and make it answer my questions.

I would sometimes feel a long, slow burning wave of ecstasy that came upon me out of the blue sky, the green trees, or across a sparkling stream. I could not make it happen. It was like a breeze that blew through me and then moved on. It seemed to say, "Follow me!" But I couldn't. And although my days were passed under the full-spectrum, 100-watt glare of reason, my nights were different. From early childhood on, just before drifting into sleep I would have visions. I would see a slow, stately kaleidoscopic progression of gorgeously colored patterns, like a private movie screened on my closed eyelids. If I walked in the woods that day, the featured attraction would be an exquisitely detailed replay of the forest floor, the intricate mosaic of fallen leaves—white oak, post oak, hickory, pine, sweetgum, and dogwood—as real as if my eyes had filmed it for later viewing.

Much later, I learned that my nocturnal visions are known as *hypnagogic hallucinations*. According to the scientific literature, they are usually very brief, highly unpleasant, and a symptom of narcolepsy. (They can also occur as a side effect of some of the serotonin reuptake inhibitors such as Effexor and Prozac.) Of course, no one goes to the doctor to complain of beautiful and prolonged nocturnal visions. Like auditory hallucinations, they only come to the attention of biomedical researchers when the hallucinations are intensely upsetting or symptomatic of an underlying disorder. With the same instinct of conformity shared by most people who hear voices, I never told a soul about mine. A survey by sleep researcher Maurice Ohayon of 13,000 European adolescents and adults found a surprisingly high prevalence rate of 25 percent for hypnagogic hallucinations, usually occurring as fleeting snapshots of the day's activities. As with auditory hallucinations, unpleasant or terrifying night visions are more likely to be associated with a sleep disorder or mental illness.

They were precious to me, those lovely nighttime visions and those fleeting daytime surges of euphoria. They hinted at some truth or reality beyond self and reason which I could not find explained in any of my textbooks. I thought that tiny tab of LSD might expand, deepen, and reveal what was hidden in those moments and by doing so, fill in the ugly blanks and gaps that I felt inside me. It was the thought of those dark corners that made me pause over the prospect of shuffling off consciousness and opening all the hatches.

Margot's acid trips seemed to be entirely positive. She, to all appearances, was a calm, mellow person, never displaying anger, self-doubt, confusion, or anxiety. I too tried hard never to show such feelings, but I certainly experienced them. I had heard about bad trips, of course. When the lid was taken off my brain, which would come out: the butterflies or the worms? No matter whether the visions were delightful or horrific, the prospect of being entirely under their sway for some hours—which might feel like centuries—ultimately seemed more claustrophobic than liberating.

I sold the tab of acid back to my roommate at a discount, yielding her a small profit. (She obviously had more worldly sophistication than me, too.) Considering the fictional and nonfictional experiences of Paul Atreides, Albert Hofmann, Wade Davis, and others, I think I made the right decision. But of course I'll never know.

CAROL HART, Ph.D., is a freelance health and science writer based in Narberth, PA, just outside of Philadelphia. She is the author of *Good Food Tastes Good: An Argument for Trusting Your Senses and Ignoring the Nutritionists* (forthcoming, SpringStreet Books) and *Secrets of Serotonin* (St. Martin's Press, 1996, with a revised and expanded second edition forthcoming in early 2008).

References

Aghajanian, G.K., G.J. Marek. "Serotonin and Hallucinogens." *Neuropsychopharmacology* 1999; 21(2S): 16S–23S.

Aghajanian, G.K., E. Sanders-Bush. "Serotonin." In: Davis KL, D. Charny, JT Coyle, C. Nemeroff, eds. *Neuropsychopharmacology: The Fifth Generation of Progress*. Philadelphia: Lippincott, Williams & Wilkins (available online at the Web site of the American College of Neuropsychopharmacology), 2002:15–35.

Balick, M.J., P.A. Cox. *Plants, People, and Culture: The Science of Ethnobotany*. New York: Scientific American Library, 1996.

Davis, W. *Shadows in the Sun: Travels to Landscapes of Spirit and Desire*. New York: Broadway Books, 1998.

Emmert, S. "Banisteriopsis caapi." Ethnobotannical Leaflets [online journal]. Summer, 1998. Southern Illinois University Carbondale http://www.siu.edu/~ebl/.

Fauman, M.A. Study Guide to DSM-IV. Washington, DC: American Psychiatric Press, 1994.

Herbert, F. *Children of Dune*. New York: Berkley Books, 1981.

———. *Dune*. New York: Berkley Books, 1977.

———. *God Emperor of Dune*. New York: Berkley Books, 1981.

———. *Heretics of Dune*. New York: Berkley Books, 1986.

Hofmann A. *LSD: My Problem Child*. 1979 (English translation, 1983). Sarasota, FL: MAPS, 2005.

Honig, A, M.A. Romme, B.J. Ensink, S.D. Escher, M.H. Pennings, M.W. deVries. "Auditory hallucinations: a comparison between patients and nonpatients." *Journal of Nervous and Mental Disease* 1998;186 (10):646 –651.

Ohayon, M.M. "Prevalence of hallucinations and their pathological associations in the general population." *Psychiatry Research* 2000; 97 (2–3):153–164.

Premont, R.T., R.R. Gainetdinov, M.G. Caron. *Following the trace of elusive amines*. Proceedings of the National Academy of the Sciences 2001; 98:9474–9475.

Snyder, S. *Drugs and the Brain*. New York: Scientific American Library, 1986.

Thomas, D.R. "5-HT5A receptors as a therapeutic target." *Pharmacology & Therapeutics* 2006; 111:707–714.

MY SECOND SIGHT: HOW TLEILAXU EYES CHANGED MY LIFE

Sergio Pistoi, Ph.D.

It has been said that the "Eyes are the mirrors of the soul." What if your eyes aren't really yours, though? Sergio Pistoi, Ph.D., shares with us history of the Tleilaxu eyes.

(From *The Panamerican Review of Science*, May 17, 2078)

A serene smile spread over the ghola's dark features. The metal eyes lifted, centered on Paul, but maintained their mechanical stare. "That is how I am called, my Lord: Hayt."

—*Dune Messiah*, FRANK HERBERT, 1969

'M A PREDICTABLE GUY. Every day I check the news on the tablecloth screen while I eat breakfast. I pick up a couple of stories that I transfer on the Netpod in case I want to watch them at work. When I have finished I change the screen's look for my wife Eloise. She likes to find that red-and-white flower design on the table. Being an old guy, I'm still fascinated by how they could fit a computer and a flexible holographic screen on a tablecloth but, hey, that's technology.

Before going out I always take a look out the window. Today is a nasty, mucky day. Big drops of rain are slowly trickling down my geranium. I told you my life is ordinary. There would be nothing special in what I do if it wasn't for one detail: I was blind. My retinas were devastated by retinitis pigmentosa, a genetic disease that has been running in my family for generations. I am one of millions of people who can see today thanks to bionic neuromorphic visual prostheses, better known as Tleilaxu eyes. My story wouldn't be worth telling if it wasn't for another detail: back in 2025, I was the first patient to receive a Tleilaxu eye. Don't search for my picture on the Wayback Machine. At that time, I asked for anonymity.

The name *Tleilaxu*, which has improperly become synonymous with bionic eyes, is actually a trademark of Tleilaxu Inc., an Indonesian company that produced the first commercial artificial retina. Chahaya Rossi, founder of the company, chose this name after the famous 1960s science-fiction saga *Dune*, in which many characters wore metallic artificial eyes produced by a secretive society called the Bene Tleilax. Real world Tleilaxu eyes have little analogy with the ones envisioned in *Dune*, except, of course, for their function. They come in dozens of models, with two basic flavors: retinal implants that substitute the retina, and bio-eyes—or full implants—that substitute the entire eye.

My new eyes arrived just in time. I was twenty and my peephole had just closed. Doctors call it tunnel vision: your visual field shrinks progressively with retinitis pigmentosa, starting from the edges until it's like you're looking through a peephole. When the peephole closes, you're blind. The Tleilaxu 1 was little more than a prototype. It had a low resolution and troubles with color, but it worked. Technically, I have been blind only for two months.

My grandfather, Primo, who also had retinitis pigmentosa, was not so lucky. After he became blind during the 1990s, he spent the rest of his life in the dark. It would take two generations after him before technology would allow blind people to see again. However, at the turn of the century, a handful of pioneers was already paving the way for Tleilaxu technology. My grandfather was a teenager when an American scientist created the forerunner of modern Tleilaxu eyes. It was exactly a century ago.

In 1978, a blind patient from New York named Jerry underwent surgery to receive the first bionic eye, pioneered by private U.S. biophysicist William Dobelle (1941–2004). Dobelle's idea was based on studies from the 1950s showing that if you stimulated the visual cortex (the image-processing part of the brain) with electrodes, you could arouse faint light sensations, called *phosphenes*, even in blind people. His system included a tiny camera (tiny for that time, of course) mounted on a pair of eyeglasses, which captured the scene and sent a signal to a computer that Jerry carried on a waist pack. Then came the bionics: the computer fed the signal to a dozen microelectrodes that a surgeon had previously inserted into Jerry's visual cortex. Jerry was probably the first wired man; you can still find old pictures of him showing bundles of wires literally hooked up to his head. The system was an impressive feat for that time but clearly it was not meant to restore full vision: all that Jerry could see were a few faint dots of light. However, with lots of training, he learned to identify some objects. Amazingly, the brain filled the gaps left by his patchy vision. Another Dobelle patient, a Canadian farmer named Jens Naumann, did even better: in 2002 he hit the headlines when he drove a car, although it was only for a few meters and on private property.

It's easy to think of the retina as a light sensor, like the one you find on your Netpod camera. Light enters the eye through the pupil; the dark, round opening in the middle of the iris, and is focused onto the bottom of the eyeball by the lens. There it encounters the retina, a half-millimeter-thick sheet that includes several layers of neurons, each with a specific function. Two types of light-sensing neurons, or photoreceptors called rods and cones, make the first layer. As their names suggest, these cells have a stretched shape. One extremity contains photopigments, chemicals that react with light, while the other is connected to the neurons of the other layers. Three different photopigments react with light at different wavelengths, allowing the retina to distinguish between colors. Rods, which are concentrated at the outer edges of the retina, require less light than cones, but have only one photopigment and therefore are unable to distinguish among colors. Because of their sensitivity, they are particularly useful at dim light. Cones are less sensitive to light than rods, but they respond faster and can distinguish between colors; they allow us to see details and colors when there is sufficient light.

However, the retina is not just a light-sensing organ. Rather, think of it as a branch office of the brain specializing in image processing. Rods and cones feed their signals to a network of neurons that includes four layers of cells called horizontal, bipolar, amacrine, and ganglion, respectively. Each of these cells in turn sends and receives input with thousands of other neurons of the retina. Ganglion cells are the last link in the chain. Their long axons (out-fibers) are bundled to form the optic nerve, which exits the eye and reaches the visual cortex of the brain. Because of the processing that takes place in the retina, the brain receives a signal only if there are significant changes in a scene, for example, if an object in the visual field is moving. When it comes to vision, it's the retina, not the brain, which takes over most of the workload: while each human retina has about 100 million photoreceptors, the optic nerve has room for only 1 million fibers. In other words, 99 percent of all visual information is filtered out before it has even left the eye.

A working artificial eye should be able to emulate, at least in part, the formidable processing that takes place in the retina. Dobelle tried to achieve that by using computer algorithms that elaborated signals before they reached the brain. However, no computer could make even with a human retina, let alone a brain, when it comes to efficiency and flexibility.

Dobelle opened the Pandora's box, but by the early 2000s his device was already considered antiquated. Many researchers deemed it too dangerous to drag electrodes and wires into the brain because of the risk of infection and seizures. Understandably, doctors preferred to experiment on sick eyes rather than on healthy brains. Therefore, many laboratories turned to systems in which microelectrodes were placed in the retina, instead of the visual cortex.

Like the Dobelle eye, these new devices featured a camera mounted on a pair of spectacles that was linked to a computer for processing the image, but the resulting signal was fed to the ganglion cells of the retina and then traveled naturally to the brain through the optic nerve. Between 2002 and 2007, researchers at the Doheny Eye Institute at the University of Southern California used such retinal implants in six blind patients, reporting that "all [of them were] able to detect light, identify objects in their environment, and even perceive motion after implantation." In the same period, German company IPP moved the concept further by putting a wireless receiver into the retina, thus eliminating the need for cables.

Retinal implants were better than the Dobelle eye and their makers relied on future improvements, but vision was still limited to faint dots of light on a dark canvas. Technological bottlenecks hampered progress: while the human retina has about 1 million ganglion cells, the chips used for the implants featured only a few hundred micro-electrodes at best. While it was possible to cram more microelectrodes into the chips, the resulting heat from the circuit would inevitably fry the ganglion cells. Also, computers that processed the images failed to emulate a real retina. While computers performed well in number-crunching applications such as weather forecasting or genetic mapping and could even beat people at chess, they could not compare to the human brain when it came to flexible tasks such as vision. Big Blue, the top supercomputer of the 2010s, was not able to recognize a human face like any newborn baby does effortlessly. Some scientists decided it was time to pursue a radically different approach. One of them was Kwabena Boahen.

If the natural retina worked so well, reasoned Boahen, why not build a silicon chip that would copy its structure? Boahen, then a young professor at Stanford University, knew that architecture was the key to explain the extraordinary performances of the retina and the brain. Unlike silicon chips which rely on a fixed wiring, neurons establish connections that adapt to a specific task. It is this architecture that allows the brain to learn and be infinitively more flexible than any silicon computer. Boahen's rationale was that if you built a processor with a structure similar to that of the retina and made it rewire itself in order to perform vision, it would behave like its natural counterpart. In 2001, together with a doctorate student, he put his idea into practice by building Visio 1, the first neuromorphic retina. It was a 3.5 x 3.3 mm silicon chip that emulated all the five layers of the human retina, replacing neurons with transistors. Phototransistors emulated cones and rods, while other types of transistors performed the functions of the different retinal neurons. Instead of physically rewiring itself (which could not be technologically possible), Visio 1 used *softwires* algorithms that virtually connected different transistors like they were real neurons. "Neuromorphic microchips, which take cues from neural structure…make it possible to develop fully implantable artificial retinas….Someday neuromorphic chips could even replicate the self-growing connections the brain uses to achieve its amazing functional capabilities," Boahen wrote in a 2006 issue of *Scientific American*. Visio 1 was a notable feat for the time: it had a resolution of about 5,700 pixels, more or less what you need to recognize a person's face, and was able to process signals almost like a real retina. The work didn't go unnoticed. Visio 1 received a lot of press, and in 2006 Boahen got the prestigious NIH Director's Pioneer Award.

Visio 1 was better than any artificial retina built before, but could not work in patients until researchers found an efficient way to connect it to the brain. Many years later, neurochips would provide a definitive solution to that problem.

The first neurochip was developed by Caltech researchers Jerome Pine and Michael Maher in 1997 and looked like an old Petri dish with a microcircuit at the bottom. As their name suggests, neurochips process information using living neurons instead of transistors. Each neurochip has hundreds or, today, millions of microscopic contacts to which the neurons are attracted. In some way a neurochip fools neurons into thinking that it is part of the brain, luring them to establish connections with its circuits. The first neurochip had room for only sixteen neurons. A few years later, Thomas DeMarse at the University of Florida went further by building a 25,000-neuron neurochip which, curiously, he used to remote-control a model aircraft. These first attempts were the beginning of a new revolution in computer science. However, it would take thirty years of trials and errors before these discoveries could benefit the blind.

The first working artificial retina, Tleilaxu 1, entered clinical trials in 2031. It was a neuromorphic retina with a resolution of 100,000 pixels, enough to recognize a face, but not to drive a vehicle without an autopilot. Conceptually, it was not very different from the Visiol models pioneered by Boahen (in fact, the Tleilaxu company had bought most of the patented technology from Stanford and Boahen himself), but the Tleilaxu included a neurochip that linked the retina and the optic nerve so that the signal could now reach the right neurons in the brain. My mother was a famous neurosurgeon, and while she denies it, I'm sure she used all her influence so that I would be included in the Tleilaxu 1 clinical trial. As it turned out, I was the first patient to receive one. A few months after surgery, my Tleilaxu retina and my brain started talking together. I was beginning to see again.

After many years, I went through surgery again for an upgrade. Now I have a Tleilaxu 5. It's still not as good as real vision—to give you an idea of what I see, think of bad bootleg copy of an holomovie—but I can drive without the help of the autopilot, and I can even appreciate the colored nuances of my flowers. At night, I actually see better than most people do. The Tleilaxu 5 is more sensitive than a natural retina and has a handy infrared mode that improves my vision in the dark. One good thing about silicon retinas is that they are free from the constraints of Mother Nature.

Early Tleilaxu models could substitute a faulty retina, but not an entire eye. Research toward full artificial eyes received an unexpected impulse in 2049, following a singular mishap that made the headlines for weeks. The third of May of that year, passengers on flight AK-290 from Singapore to New York were peacefully watching in-flight movies and news on their personal holographic spectacles when, following a sudden drop in pressure (investigators later found that a faulty safety circuit was to blame), all holoscreens on board exploded simultaneously, destroying the eyes of 531 people with their highly corrosive plasma. The accident, one of the strangest in the entire history of aviation, caused an unprecedented emotional wave in the public. Millions of people donated money to research, starting a race to build new prostheses that would substitute an entire eye. After ten years, such devices called full implants or B-Tleilaxu became available to people who lost their eyes because of accidents or cancer. They are made by combining a Tleilaxu silicon retina with a bionic eyeball, obtained by growing stem cells on biocompatible scaffolds, complete with the six orbital muscles that control eye movements. The iris and lenses, which cannot currently be obtained with stem cells, are made of synthetic materials. B-Tleilaxu eyes look almost natural, but since they move slowly and they have a fixed pupil, they give their wearer a sort of glacial, mechanical stare, which reminds one of *Dune*'s characters.

It's still raining when I come back from work. I often think that I am lucky to have good health insurance: Tleilaxu eyes, especially the new models, are too expensive for most people. Researchers agree that biological retinas made in a laboratory from neural stem cells would be a better and cheaper alternative to Tleilaxu eyes. Sadly, research on neural stem cells is practically banned in many countries, including ours, because of the pressure from religious groups. It's amazing that at the end of the twenty-first century so many people still believe that it's evil to use these cells. Man can't unravel the mystery of the soul, they say.

I will stop by to pick up Eloise and we'll go out for dinner. I guess we will argue once more about the holiday plans. She likes to visit new places, while I prefer to come back to our house in Greece, as usual. Many years ago, when I was still blind, I made a list of all the beautiful places in the world that I would visit if I could see again. It was just a way of steeling myself. To be true, I have always detested traveling. I don't like changes. I told you, I'm a predictable guy.

SERGIO PISTOI, Ph.D., started his career as a molecular biologist. Soon after his Ph.D. in 1994, a radiation accident in his lab turned him into an evil science-writing superhero. He was an intern at Scientific American in New York and a stringer for Reuters Health. His credits include *Scientific American*, *New Scientist*, *Nature*, and many Italian print and radio outlets. He is also a consultant for research planning and portfolio management. He is a member of the National Association of Science Writers NASW and the European Union of Science Writer's Associations. He hides in Tuscany, Italy, with a fake identity. He can be found at www.greedybrain.com.

THE BIOLOGY OF THE SANDWORM

Sibylle Hechtel, Ph.D.

The Bene Gesserit have their order. The navigators have their Guild. The Imperium has the Great Houses. These three orders form a triumvirate that controls the destiny of all Humanity in the Duniverse. Yet all would be completely at the mercy of the Arrakean sandworms, should they form a union. The Shai-Hulud make the spice, they make the Navigators who they are, they make the Bene Gesserit who they are, they indirectly make the Fremen who they are, and they even make Arrakis what it is. Could such a creature even exist, however? Biologist Sibylle Hechtel, Ph.D., responds to that very question.

Truth is stranger than fiction, but it is because Fiction is obliged to stick to possibilities; Truth isn't.

—MARK TWAIN

Truth is stranger than fiction; fiction has to make sense.

—LEO ROSTEN

SHAI-HULUD'S NAME strikes fear in the hearts of Dune's people.

The giant sandworm can swallow a harvesting machine, including its workforce of twenty-six men, in one gulp. Workers who harvest the priceless melange employ "carry-all"

29

helicopters to swoop down from the sky and whisk them away before the dreaded sandworm can eat the harvesting machines (workers included). On Paul's first trip into the desert, he watches from the helicopter as a worm takes a crawler:

> A gigantic sand whirlpool began forming....Sand and dust filled the air for hundreds of meters around it...a wide hole emerged from the sand. Sunlight flashed from glistening white spokes within it. The hole's diameter was at least twice the length of the crawler...the machine slid into that opening...the hole pulled back. (*Dune* 123)

Worms range in size from a small specimen, 110 meters long and 22 meters in diameter, to medium worms at about 200 meters long, to the biggest worms at over 400 meters long with an 80-meter diameter mouth. Herbert leaves much of the worm's biology vague, such as whether it is a vertebrate, how it moves or what it eats, and instead focuses on the worm's behavior and actions and their effect on Dune's population.

Herbert never describes precisely how the worm moves, only that it looks like a fish that "swims" just under the surface. He frequently describes the worm's motion in sand as "a cresting of sand," or mentions the "burrow mound of a worm" (*Dune* 414). The worm primarily comes above the surface when it's eating a 'thopter or crawler, or when the Fremen catch one and put their hooks in its scales to drive it up out of the sand. He describes the worm as eating harvesting machinery and ornithopters, complete with occupants. What it may eat when underground, far from human habitation, Herbert leaves to our imagination. Of the sandworm's lifecycle, Herbert mentions:

> The circular relationship: little maker to shai-hulud; shai-hulud to scatter the spice upon which fed...sand plankton; the sand plankton growing, burrowing, becoming little makers [sandtrout]. (*Dune* 497)

He describes sand plankton that grow into sandtrout, some of which grow and eventually metamorphose into new sandworms. But he does not explain whether the sandworm lays eggs to create the sand plankton, whether there are male and female worms, or how reproduction occurs.

I will discuss which of the sandworm's characteristics seem plausible in terms of what we know of terrestrial life, which attributes could not occur by any mechanism we know of on Earth, and how the sandworm could instead carry out certain functions in light of terrestrial biology. I'll speculate on certain factors that Herbert leaves vague, such as what do all those sandworms eat, and where does the food grow?

Why don't we see large sandworms roaming the Sahara or Chile's Atacama Desert, the driest desert in the world with places where no rain has ever been recorded? When we explore other planets, might we find creatures resembling Dune's sandworm? Possibly.

Sandworm, Sandtrout, and Spice

Let's look at the worm's life cycle and ecology to see how closely Shai-hulud fits with terrestrial animals. Liet-Kynes, His Imperial Majesty's Planetologist and Arrakis's planetary ecologist, predicted the existence of an underground organism like the sandtrout, because some organism must produce oxygen. Since very, very few plants grow above ground to generate oxygen, there must be some oxygen-producing life form underground. (Plants produce most of Earth's oxygen. Before the evolution of photosynthetic plants, Earth's atmosphere consisted of carbon dioxide, hydrogen, nitrogen, and methane, with little or no free oxygen.)

Kynes says:

> How strange that so few people ever looked up from the spice long enough to wonder at the near-ideal nitrogen-oxygen-CO_2 balance…in the absence of plant cover…something occupies that gap. I knew the little maker was there, deep in the sand, long before I ever saw it. (*Dune* 274)

Terrestrial life requires carbon, water, and an energy source. Photosynthetic plants take in carbon dioxide and water, using energy from the sun (photons) to split water and convert carbon dioxide (CO_2) into carbohydrates—a long chain composed of carbon, hydrogen, and oxygen. They release waste oxygen (from the carbon dioxide, which has two oxygens for one carbon) into the atmosphere, providing us with the air we breathe. Scientists postulate that the early atmosphere consisted primarily of CO_2 and methane (CH_4) until the advent of photosynthetic plants to convert CO_2 into free oxygen.

If the sandtrout generates oxygen, what would it use as an energy source? On Earth, chemoautotrophic bacteria use chemicals, usually hydrogen sulfide (H_2S), to supply the energy to synthesize carbohydrates from CO_2, similar to plants, and release oxygen. Since the sandtrout live deep underground, away from the sun and any means of photosynthesis, they would have to use chemicals, most likely H_2S, to supply their energy requirements. H_2S serves as an energy source for chemoautotrophic bacteria on Earth in deep-sea hydrothermal vents communities, some hot springs, and some caves. Since H_2S often escapes from underground vents formed as a result of volcanic activity, it's the most likely source of subterranean energy on Dune.

What is the relationship between the sandworm and spice?

The pre-spice mass had accumulated enough water and organic matter from the little makers....A gigantic bubble of Carbon dioxide was forming deep in the sand. (*Dune* 277)

Herbert describes the "little maker," or "sandtrout," as "a sandswimmer that blocked off water into fertile pockets within the porous lower strata" (*Dune* 497).

It's plausible that the sandtrout themselves produce melange. Alternatively, they may tend an organism underground, most likely a fungus (which requires no light), that synthesizes melange. The sandtrout could prepare a suitable environment for the spice-producing organism by segregating and storing water. Herbert mentions no spice-producing organism other than the sandworm. But in terrestrial biology, plants, bacteria, and fungi produce the majority of exotic compounds, and we should consider this possibility.

We see "gardening" activities among ants. Leaf-cutting ants cut growing tree leaves and drag them to an underground growth chamber in their nest, keep it moist, and cultivate fungi on the leaves. The ants eat this fungus, which grows only underground in their nest. A second symbiotic bacterium grows on the ants and secretes chemicals, which protect the fungus from mold. Here the ants use antimicrobials to protect their harvest like we use insecticides in our fields ("Fungus"; "Leafcutter Ant").

Plants synthesize numerous different "secondary chemicals" as a defense against parasites or herbivores. These thousands of secondary chemicals, so called because they're not essential for the primary metabolism involved in growth, photosynthesis, and structural support, serve purely as a defense mechanism. Secondary plant compounds comprise alkaloids, cardiac glycosides, cyanide-containing compounds that can release cyanide when an insect tries to eat the plant, non-protein amino acids, and many others. Some well-known chemicals that plants synthesize to keep insects from eating them (not for our benefit) include the alkaloids such as cannabis, cocaine, opium, nicotine, caffeine, and digitalis (in foxglove). Other secondary plant chemicals work as antibiotics or to alleviate pain, like salicylic acid in willows (aspirin) or penicillin (a mold). Possibly either the sandtrout or the underground fungi produce spice to protect them against bacteria.

Other secondary compounds exist in marine snails and sponges (conotoxins), and Amazonian Indians have long used poison from toads (bufotoxin) to coat their lethal arrowheads. Chinese folk remedies prescribe a related extract from the skin of Asian toads. They ascribe life-prolonging attributes to these remedies, similar to the purported life-extending properties of spice, but in the case of Earth's known plants with secondary chemicals, the life-extending effects have yet to be found. Existing terrestrial plants and animals both produce chemicals with activity similar to spice, in that they cause hallucinations and have other mind-altering properties.

The sandtrout block off water in underground pockets and could cultivate fungi deep below the surface, where they are safe from the arid, drying winds and from dehydration by the fierce sun. Herbert never specifies what the sand plankton and sandtrout eat, except possibly spice. But if he says that the sandworm produces the spice, how can the sandworm/plankton/trout simultaneously produce and subsist on melange? It makes more sense, in view of terrestrial biology, that another organism, like a fungus, grows underground and produces both oxygen and spice. The sandtrout could then consume the fungi, much like terrestrial leaf-cutting ants subsist on fungi in their underground nests, and either synthesize spice themselves or get it from the fungi. Hallucinogenic chemicals abound in mushrooms, such as psilocybin and the Amanita mushroom (*Amanita muscaria*).

The diet of Shai-hulud is intimately connected with its life cycle. Let's review Kynes's description of both:

Now they had the circular relationship: little maker to pre-spice mass; little maker to shai-hulud; shai-hulud to scatter the spice upon which fed microscopic creatures called sand plankton; the sand plankton, food for shai-hulud, growing, burrowing, becoming little makers [sand-trout]. (*Dune* 497)

The sandtrout is a key element both in Shai-hulud's continued survival and in spice production. The sandtrout encapsulates water pockets underground where fungi can grow; these, in turn, nourish both the sand plankton and more sandtrout. Herbert describes leathery scraps of material found with the spice mass after a blow. When the surviving sandtrout burrow down deep to encapsulate more water pockets from the residual water drops, the remains of the dead sandtrout can serve as additional food for the fungi.

We find similar ecologies deep beneath the sea in the ocean floor of the Galapagos Rift (more than 8,000 feet deep). Giant red tube worms (Vestimentifera) and various bivalves grow adjacent to hydrothermal vents. The volcanically active oceanic ridges create molten rock, which heats seawater, often to 700 degrees. These underwater "smoking chimneys" emit hot water, loaded with metals and dissolved sulfide, thus providing two of the main requirements for life—liquid water and an energy source.

In these sea-bottom communities, sulfide-oxidizing bacteria grow on rocks as mats and, in turn, limpets, clams, and mussels graze on the bacteria. However, many vent animals live with symbiotic bacteria. Tube worms have a specialized organ, the *trophosome*, with chambers that contain sulfide-oxidizing bacteria. The bacteria live in the tube worm's trophosome, and the worms digest some of their symbiotic bacteria as a food source. Some bivalve mollusks have symbiotic sulfur-oxidizing bacteria in the gills ("Hot Vents").

Imagine the sandtrout community as a combination of leaf-cutting ant nest and hydrothermal vent community. In underground caves or caverns, a vent or *fumarole* emits hot water laced with sulfur and minerals. The sandtrout sequester this water into chambers and pools. Alongside, the surviving sandtrout drag the scraps of their deceased brethren down to the nest where, on one side, bacteria consume sulfide and minerals and grow into large bacterial mats. In other chambers, spice-producing fungi grow on dead sandtrout scraps and dead bacterial mats. The sand plankton and sandtrout then devour some of the living fungi and bacterial mats.

Reproduction and Dispersal

Let's look again at Kynes's description of the life cycle:

> They had the circular relationship: little maker to shai-hulud; shai-hulud to scatter the spice upon which fed . . . sand plankton; the sand plankton growing, burrowing, becoming little makers [sandtrout]. (*Dune* 497)

Probably the little maker (sandtrout), which collects and stores water and cultivates fungi, exists as the most abundant stage in terms of biomass. Initially, when sandtrout or sand plankton colonize a new cave, the sandtrout block off the water into pockets where the hypothetical fungi or algae grow. The sand plankton could then feed on the speculative underground bacteria or fungi that grow near vents. The sandtrout could grow and reproduce for many generations by budding, like some marine invertebrates including jellyfish, or by fragmentation, like ribbon worms.

Many organisms—bacteria, fungi, plants, and lizards—reproduce asexually (parthenogenesis) as genetically identical clones until the environment changes or deteriorates, which signals a switch to sexual reproduction. Exchanging genes between genetically different individuals creates new gene combinations, which might better adapt to the new, different environment. Some animals even switch sex to accomplish this, such as certain fish and amphibians (Rice).

The sandtrout could survive indefinitely as a clone form until some environmental change—like, perhaps, the accumulation of carbon dioxide resulting from the rampant growth of fungi and sandtrout (the pre-spice mass)—triggers sexual reproduction. If genetically different individual sandtrout mate, then they will lay eggs that hatch into individual sand plankton with novel gene combinations. When the pre-spice mass blows as a result of carbon dioxide accumulation (from the respiration of millions of sandtrout), the force would scatter the sand plankton (and spice) widely.

The giant sandworm comes to the spice blow, not necessarily only to eat spice, but also to help disperse its offspring to a new environment where they can colonize a new cave and survive to grow. Sand plankton and sandtrout cannot disperse across the desert over long distances unaided. When the spice mass blows, Shai-hulud scoops up as much of the plankton-laden sand as possible and takes its offspring to other caves. After colonization of a new vent and further reproduction, the sandtrout then lay numerous eggs, which hatch into tiny sand plankton that feed on spice, fungi, and sulfur-reducing bacteria. The sand plankton eventually grow larger and become sandtrout. A few of the (largest and most ecologically successful) sandtrout could grow into another sandworm:

> The few survivors entered a semi-dormant cyst-hibernation to emerge in six years as small (about three meters long) sandworms…only a few avoided their larger brothers and pre-spice water pockets to emerge into maturity as the giant Shai-hulud. (*Dune* 497)

Insects, similarly, spend the majority of their life span in an immature feeding stage. Caterpillars, which are juvenile butterflies or moths, feed and grow all season and then molt and pupate, usually in a cocoon or enclosed burrow, during winter. A pupa doesn't eat and, instead, remodels its body, using energy supplies accumulated during the feeding larval stage. It undergoes metamorphosis inside the protected pupal case over the inhospitable season, while food is unavailable, and emerges in spring as a butterfly or moth. Similarly, the sandtrout could form a cyst, hibernate, and emerge as a small sandworm. The smaller sandtrout would need to grow much larger by eating sufficient spice, fungi, bacteria, or other sandtrout to acquire enough biomass that it could turn into a three-meter sandworm. Remaining in hibernation, or a pupal stage, for six years is not unusual: seventeen-year cicadas, sometimes referred to as seventeen-year locusts, pupate and remain inactive for thirteen to seventeen years (Cicada Mania).

Biologists ascribe insects' remarkable success, in part, to metamorphosis. This dramatic change of shape allows the adult more mobile form to exploit a very different environment from the juvenile form. For instance, caterpillars live in a narrowly circumscribed locale, feeding on leaves or fruit. In contrast, adults, like the Monarch butterflies, sip nectar from flowers and migrate almost a thousand miles from their wintering areas in Mexico to summer feeding grounds. One butterfly flew a 1,870-mile route from Ontario to Mexico in four months ("Voyagers"; "Metamorphosis").

Not only insects metamorphose, changing form dramatically, but also marine animals. Cnidarians, which include jellyfish and the Portuguese man-of-war, alternate between two body shapes in their life cycles: a polyp, immobile stage, and a more mobile reproductive stage, the medusa (swimmer that looks like a jellyfish). Polyps often grow in colonies, reproduce asexually by budding (growing shoots, like plants), and look a little like sea anemones with a mouth on top, surrounded by tentacles. The medusa form, which we know as a jellyfish, grows much larger than the polyps (like the sandworm is much larger than the sandtrout), the largest jellyfish measuring more than two meters in diameter with tentacles about thirty meters long.

But sandworms aren't insects or jellyfish!

A third group of organisms that undergo metamorphosis is amphibians—frogs and toads. Adult frogs lay eggs, in or near water, that hatch into tadpoles, which have gills and tails, eat plants, look a bit like fish, and live in lakes or streams. During metamorphosis to the adult form, the gills disappear; they absorb the tail and grow legs, and shift from eating plants to adopting a carnivorous diet. Not surprisingly, the amphibians that exploit either land or water in the very different adult and juvenile forms became the first successful vertebrates on land. Thus the sandworm's metamorphosis from sand plankton to sandtrout to sandworm differs only in magnitude from the terrestrial model of metamorphosis. We see three highly successful kinds of organisms: insects, which go from land herbivore to airborne butterfly; Cnidarians, which go from sessile polyp to mobile, much larger medusa; and frogs, which transform from a vegetarian aquatic tadpole to a carnivorous frog. The sandworm, which starts as grazing plankton, grows into a vegetarian sandtrout, and then metamorphoses into the giant worm, fits in well with this model.

The adult giant sandworm accomplishes one more essential function—dispersal and colonization of new habitats. The small sandtrout, and even more so, microscopic sand plankton, can't migrate long distances across sere desert sands. I would assume that on Dune, like on Earth, fresh volcanic rifts or fumaroles periodically appear in caves deep beneath the surface ("Fumarole").

Assuming that Shai-hulud can smell the sulfurous gases emitted deep below the sand when it travels across the desert, it can head toward the new source of energy, burrow down, and leave behind some sand plankton or sandtrout to begin a new colony at the recently created volcanic vent.

One question that terrestrial scientists still debate is how marine organisms find and colonize the widely scattered ephemeral vents on the ocean floor. Colonizing newly formed volcanic vents below the sea could occur in three ways: (1) as a result of organisms drifting with ocean currents from an existing hydrothermal vent colony to a newly formed vent, (2) by organisms drifting down from surface waters where they floated on mats of floating seaweed, or (3) by organisms that were growing on a large fish or whale (barnacles sometimes grow on them) and that drifted down to the bottom when the fish or whale died. None of these methods of dispersal would readily work on Dune's sandy deserts. Caterpillars have an adult form, the butterfly, which finds the widespread, patchily distributed clumps of plants or flowers its offspring prefer to eat, and then lays its eggs on those plants, ensuring that the next generation finds enough of the right food. Similarly, sandworms can distribute sand plankton to newly formed vents in underground caves. Herbert said, "Shai-hulud to scatter the spice."

Marine scientists, however, remained unaware of ocean-floor hydrothermal vent communities until their discovery by the 1977 expedition to the Galapagos Rift. Herbert and Kynes would have been unaware of communities based around sulfur-gas emitting vents. Hence he would not have suggested that the sandworm would smell the sulfurous vents and head there to scatter its load of spice and sand plankton, much like turtles lay their eggs in fertile habitats, insects lay eggs on a juicy carcass, and salmon head upriver to spawn. Animals seek the best potential habitat for their offspring before they lay eggs or spawn.

The Size of Sandworms

Next, let's look at how close terrestrial animals, both present and past, come in size. Dune's sandworms range from medium worms of about 200 meters long, to half a league for the biggest, with an 80-meter-diameter mouth with which it can swallow a 120-by-40 meter harvester in one gulp. A league measured 1.5 Roman miles in ancient Rome and about 3.25 kilometers to 4.68 kilometers in France, so we're looking at several kilometers in length!

What terrestrial animal does Shai-hulud most resemble? An earthworm, in the group "annelid" worms, which lives mostly in water or moist habitats. These and related worms lack an integument, or outer covering that resists desiccation—a must on a harsh desert world with Dune's fierce winds.

Some arthropods, such as centipedes and millipedes, possess a hard exoskeleton, which covers their entire body and appendages. The exoskeleton prevents water loss, supports the tissue, and provides a place for muscles to attach. However, any animal with an exoskeleton will eventually outgrow its shell. It must then shed its old shell (molt) and wait for its new shell to harden. While this process is going on, the hapless worm would lose vast quantities of water from its body and remain vulnerable to high winds.

And last, an exoskeleton limits the ultimate size an animal can attain. A giant spider, like Aragog in *Harry Potter and the Chamber of Secrets*, would never grow large enough to menace Harry and his friends because of the limits an exoskeleton imposes on ultimate size. Nor will giant alien ants ever menace Earth. The sandworm cannot attain the size that Herbert postulates with an exoskeleton. In order to reach anywhere near the magnitude that Herbert suggests, the sandworm must be a vertebrate and more of a "sand snake." At one point, Herbert mentions its "scales," suggesting a more reptilian or snake-like creature.

> Paul glanced down at the scaled ring surface on which they stood. . . . Bottom scales grew larger, heavier, smoother. Top scales could be told by size alone. (*Dune* 403)

Let's consider the largest animals—extinct giant dinosaurs and modern-day whales. Both are vertebrates, with a support structure—the skeleton—on the inside. This internal skeleton, comprised of living tissue, grows larger as the animal grows and, thus, would not limit the ultimate size a sandworm could attain.

Dinosaurs, the largest of land animals, ranged from the twenty-meter long *Apatosaurus*, about 5 meters tall and weighing as much as five adult elephants, to the even longer *Diplodocus*, at almost 30 meters long. The giant *Brachiosaurus* and *Titanosaurus*, which may have been longer than thirty-three meters, were the largest animals that ever lived on land. Still, Shai-hulud dwarfs even the largest land animals.

What about whales? Blue whales are the largest animal ever found—bigger even than the dinosaurs. The largest whale weighed 171,000 kilograms and measured over 27 meters long. The longest whale was more than 33 meters long, but still a pygmy when compared with the sandworm.

Are there theoretical or practical limits to how large an animal can grow? In D'Arcy Wentworth Thompson's monograph *On Size and Form*, he discusses the limits that physics places on specific animal and plant structures. Both animals and plants need to accomplish several tasks: they need to transport nutrients in, waste products out, protect themselves from predation, and successfully reproduce. The nutrients they require include food, water, and oxygen in the case of animals and carbon dioxide for plants. Waste products they need to remove, in addition to feces and urine, include waste heat generated by metabolic processes. When we use our muscles to exercise, such as when an animal pursues its next meal or runs to escape from a predator, muscles generate heat. In very cold weather, when our body temperature drops, we shiver to generate more heat, and in hot temperatures we sweat to get rid of excess heat. Heatstroke occurs when the ambient temperature and humidity are so high that we cannot remove the excess heat generated by activity, and our body temperature rises to dangerous levels.

The surface area of a body, such as a cylinder, cube, or sphere, increases as the square of the radius (r^2) increases, whereas the volume of that same body increases as the cube of the radius (r^3) increases, which means that as an animal's size increases, the volume increases much faster than the surface area. As animals engage in strenuous activity, they generate heat as a function of the musculature and hence of volume. Normally the surface enables animals to efficiently lose heat, but with a larger volume-to-surface area ratio, larger animals will have a harder time disposing of extra heat.

Medium to large mammals, from dogs to humans to elephants, have various adaptations to help with thermoregulation. In order to dissipate excess heat, they pant (dogs) or sweat (humans). Biologists hypothesize that the elephant's large ears serve as a type of radiator to facilitate heat loss. Also, herbivorous elephants don't need to chase their food and, thus, generate less heat than predators like wolves! Anthropologists suggest that one factor contributing to man's success as a hunter was his bare skin and sweat glands that allowed him to pursue his prey even on the hot savannah without succumbing to heat prostration.

How, then, would the much larger sandworm, with a scaly, water-impermeable or nearly impermeable integument, rid itself of excess heat generated from chasing prey? A time-honored method consists of evaporative cooling, allowing sweat on the body to help dissipate heat by the transformation of water to steam (water vapor; a phase change which requires an input of heat energy). However, using sweat as a method of cooling on Dune is clearly prohibitive. Perhaps this helps explain, in part, the "hot chemical furnaces churning inside the worm"—the sandworm cannot lose heat by any conventional terrestrial method, such as sweating, panting, or evapotranspiration, and instead accumulates the heat energy, generated by muscles during periods of high activity, and uses that heat to drive chemical reactions inside its body.

Structural Limits

Physics further limits size structurally, such as the ultimate potential height to which trees can grow. Above a certain height, a tree will bend due to its own weight. An animal's skeleton resembles, in principle, a beam supported at either end. Beams carrying no additional weight sag downward proportionally to the square of their length and cross-sectional size. Given two similar beams, one 5 centimeters long and the other two meters long, the longer sags 1,300 times as much as the shorter beam.

Applying this to terrestrial animals, we find that, as an animal's size increases, its skeleton gets bulkier and heavier. Bones, as a percent of body weight, compose:

8 percent of the body of a mouse or songbird
13 to 14 percent of a goose or dog
17 to 18 percent of a man
In theory, up to 40 percent of a small sandworm. (Thompson 20)

To build a larger land animal, we need to use harder and stronger structural materials. The sandworm must have bones with a metal matrix instead of rock, like terrestrial animals (calcium, a chief constituent of bones, is heavy and breaks relatively easily). Imagine instead bones made of titanium, steel, aluminum, or a chrome-molybdenum alloy. These bones would be significantly stronger, thinner, and possess more tensile strength than our standard stony bones.

To support a 200,000 kilogram weight with a 100-meter-long bone requires a bone with a radius 0.5 meters, which would weigh about 155,000 kilograms, or more than three-quarters of the worm's total body weight. Titanium "bone" only requires a radius of 0.3 meters, but titanium is heavier, so the weight of the titanium "bone" is still 136,000 kilograms. A "bone" made of carbon nanotubes would be only 0.17 meters in radius and weigh about 24,000 kilograms ("Carbon Nanotubes"; Dalton).

For a very sophisticated worm that can grow carbon nanotubes, the "bone" is 11 percent of the weight. For a titanium bone, it would be 40 percent, but then it's unclear if the bone could support itself. Titanium bones would account for the sandworm's rapacious appetite for harvesting machines! The complete human skeleton is replaced every ten years. For a worm that massive to replace its bone cells regularly, it would need to eat quite a few harvesters annually to meet its recommended daily requirement of heavy metals.

Dune's sandworm is larger, by several orders of magnitude, than any terrestrial animals past or present. How does the sandworm compare in size with Earth's largest plants? When Kynes crawled across the dunes, cast out by the Harkonnens without a stillsuit, he smelled a pre-spice mass in a pocket beneath him and thought of water:

He imagined it now—sealed off in a strata of porous rock by the leathery half-plant, half-animal little makers. (*Dune* 273)

He refers to the sandtrout, the immature form of the sandworm, as half-plant, perhaps to emphasize the unique and confusing nature of the beast. Let's investigate how large plants can grow: the world's largest tree (by volume) is the General Sherman Tree in Sequoia National Park, slightly over 1,487 cubic meters and 84 meters tall.

The world's tallest tree was the 112-meter-tall Stratosphere Giant until it was replaced by a new candidate, a redwood in California's Redwood National Park. According to Professor Steve Sillett of Humboldt State University, the record-setting tree, named Hyperion, measures 115 meters tall. Researchers exploring in the Redwood National Parks discovered two other redwoods taller than the Stratosphere Giant. Still, even at 115 meters, the tree is dwarfed by the worm.

Many biologists consider the world's largest organisms to be either the giant "mushrooms" or aspen clones. Michigan biologists discovered giant fungal clones that cover an area of forty acres. Another group of scientists found a 1,500-acre fungus in Washington. We have no weight yet for this super-organism, which may be the world's largest organism in area, but is not the largest in mass. Its discoverers guess that it probably weighs under 375,000 kilograms, less than a giant sequoia, which can weigh up to 2 million kilograms.

A fungus, *Armillaria ostoyae*, covered 600 hectares in Washington state. Mycology experts thought that if a monster Armillaria grew in Washington, then one as large could be causing trees to die in the Malheur National Forest of Oregon's Blue Mountains. They found one that they estimated covered more than 2,200 acres (890 hectares) and was at least 2,400 years old.

Some scientists consider quaking aspen to be the world's largest organisms: a stand of thousands of aspens is actually one single organism, which shares the same root system and has identical genes. The largest known aspen clone in Utah's Wasatch Mountains contains 47,000 trees that cover about 106 acres (Grant). Scientists estimated that it weighs over 5.9 million kilograms, which makes it three times heavier than the General Sherman Tree sequoia tree and dwarfs even the giant fungi. While terrestrial animals come nowhere near Dune's sandworm in size, some plants come close. Perhaps that is why Herbert calls the sandtrout half-plant—only plants can grow as large as the sandworm!

How Does It Move?

When we look at how animals disperse to colonize new habitats, we need to look at locomotion—how do these organisms move? Do they walk, crawl, swim, or fly? In the case of Shai-hulud, Herbert remains reticent about how the worm moves. (If I had to explain how a one-mile long creature walks or crawls, I too would be as vague as possible!)

Instead Herbert uses marine analogies. Kynes says one must avoid "tidal dust basins" which occur when:

> Certain depressions in the desert have filled with dust over the centuries. Some are so vast they have currents and tides. (*Dune* 117)

When they saw the worm, there:

> Came an elongated mount-in-motion—a cresting of sand. It reminded Paul of the way a big fish disturbed the water when swimming just under the surface. (*Dune* 118)

Later, Herbert says:

> The worm came on like some great sandfish, cresting the surface, its rings rippling and twisting. (*Dune* 414)

Herbert describes the sandtrout as a "sandswimmer"—continuing in the vein of comparing Dune's sands to the sea. He avoids an insoluble physical problem in portraying the sands of Dune as having properties similar to water. Earthworms burrow through the soil, but at an achingly slow pace. They actually "eat" their way through the dirt, in a way, by turning their mouth into a type of wedge.

The remarkable speed that Herbert ascribes to the giant sandworm wouldn't be possible underground in a terrestrial type of medium. Instead, he gives Dune's desert sands properties similar to water.

This solves further structural problems—if the sandworm lives in a buoyant, waterlike medium, then this medium helps support its mass and the sandworm can more easily grow to a larger size. D'Arcy Thompson mentions that for an animal immersed in water, its weight is counterpoised with the water around it and there no longer remains as great a physical barrier to indefinite growth. He further points out that in an aquatic animal, the larger it grows, the faster it goes. Its energy (for locomotion) depends on muscle mass (a function of volume), but its motion through water is opposed only by friction (a function of surface area), not by gravity, as on land.

However, if we assume that Dune's sand acts like water, then we are faced with a contradiction, since the men of Dune can no longer walk the sands.

In conclusion, Earth could not support a terrestrial animal the size that Herbert ascribes to the sandworm. The largest dinosaurs, somewhat longer than 33 meters, were the largest animals to live on land. Even the largest whales are barely over 33 meters, as compared to the 200-meter sandworms. An aquatic animal could grow larger than a terrestrial one, since the water's buoyancy helps support its weight. Also, in an aquatic medium it's easier for animals to swim than walk; heat exchange is less of a problem since the worm can dispose of its excess metabolic heat directly into the water, which has a much higher heat capacity than does the land. (Heat capacity is a chemical property of matter. Water can absorb and store much more heat than land, which is why seaside climates are generally much milder than continental ones.)

The question of what produces the spice and generates oxygen remains vexing. On Earth, all oxygen production stems from plants, single-celled organisms like algae, and photosynthetic or chemosynthetic bacteria. No animal produces oxygen. Herbert states in the appendix that:

> Even Shai-hulud had a place in the charts…his inner digestive "factory," with its enormous concentrations of aldehydes and acids, was a giant source of oxygen. A medium worm (about 200 meters long) discharged into the atmosphere as much oxygen as ten square kilometers of green growing photosynthesis surface. (*Dune* 499)

On Earth, the only mechanism for a sandworm to generate oxygen would be if chemoautotrophic bacteria inhabited its gut, much like sulfide-oxidizing bacteria that live in the trophosome of tube worms at deep-sea vents ("Hot Vents").

Thus, when we look for Shai-hulud on different planets, we may not find a sandworm in the Sahara or on Mars, but perhaps on Jupiter's moon Europa (Irion), where it could swim the subsurface seas like some great Loch Ness monster.

SIBYLLE HECHTEL received her Ph.D. in biology from the University of California, Irvine. She taught the Biology of Aging at the University of Michigan while researching mitochondrial DNA evolution and later worked as a faculty research fellow at Caltech studying repetitive DNA. After several years of working sixty to seventy hours weekly in labs with no windows, she quit academics to work as a writer. Her work has been published in *New Scientist, Red Herring, Reuters Health*, and others. She wrote a book on rock climbing due out in 2007.

References

"Biggest Living Organism. Fungus, *Armillaria ostoyae*." *Extreme Science.* 2006. http://www.extremescience.com/biggestlivingthing.htm

"Carbon nanotube." Wikipedia, the Free Encyclopedia. http://en.wikipedia.org/wiki/Carbon_nanotube#Strength

Cicada Mania. http://www.cicadamania.com/cicadas/

Dalton, Aaron. "Nanotubes May Heal Broken Bones." *Wired.* Aug. 15, 2005. http://www.wired.com/medtech/health/news/2005/08/68512

"Fumarole." Wikipedia, the Free Encyclopedia. http://en.wikipedia.org/wiki/Fumarole

"Fungus." Thinking Fountain. *Science Learning Network.* http://www.thinkingfountain.org/f/fungus/fungus.html

Grant, Michael. "The Trembling Giant." *Discover.* 1 Oct. 1993. http://discovermagazine.com/1993/oct/thetremblinggian285

Herbert, Frank. *Dune.* New York: Ace Books, 1990.

"Hot vents." *The Biological Sciences: State University of New York at Stony Brook.* http://life.bio.sunysb.edu/marinebio/hotvent.html

Irion, Robert. "Like Alaska, Like Europa." *Discover.* 1 May 2002. http://discovermagazine.com/2002/may/cover

"Leafcutter Ant." Wikipedia, the Free Encyclopedia. http://en.wikipedia.org/wiki/Leafcutter_ant

"Metamorphosis—A Remarkable Change." *Australian Museum Online.* 2004. http://www.amonline.net.au/insects/insects/metamorphosis.htm http://www.wired.com/medtech/health/news/2005/08/68512

"Monarch Migration." *PBS Online.* Nature. Alien Empire. Voyagers. http://www.pbs.org/wnet/nature/alienempire/voyagers.html

Rice, Aaron N. "The Physiology of Sex-Change in Coral Reef Fish." *Davidson College Homepage*. 1999.
http://www.bio.davidson.edu/Courses/anphys/1999/Rice/Rice.htm

Thompson, D'Arcy. *On Growth and Form*. Abridged ed. Cambridge University Press, 1971.

THE DUNES OF DUNE: THE PLANETOLOGY OF ARRAKIS

Ralph D. Lorenz, Ph.D.

As a book title, Planet of the Sandworms *simply fails to convey the same majesty as* Dune. *Sand dune expert Ralph D. Lorenz, Ph.D., discusses the science behind what makes Arrakis...* Dune.

UNE—it is striking that a mere four letters, the name of a simple and not uncommon landform, should be so evocative. Alternatives like lake or hill somehow lack the exotic majesty needed to act as the name of a book and a planet.

Ironically, the book said little about the dunes themselves—the characters and intrigue were the focus, and rightly so. However, Herbert's *Dune* world Arrakis is interesting to consider in a planetary context, since other worlds in our own Solar System have intriguing differences and similarities with Arrakis. Nowhere, though, is dominated by dunes to the same extent—to quote Baron Valdimir Harkonnen, "Observe closely, Piter, and you, too, Feyd-Rautha, my darling: from 60 degrees north to 70 degrees south—these exquisite ripples. Their coloring: does it not remind you of sweet caramels? And nowhere do you see blue of lakes or rivers or seas. And these lovely polar caps—so small. Could anyone mistake this place? Arrakis! Truly unique" (*Dune* 14).

Much of what is known about dunes on Earth was systematized by the British scientist Ralph Bagnold in his 1941 classic *Physics of Wind-Blown Sand and Desert Dunes.* Bagnold, an officer in the British army as well as a physicist and desert explorer, played a major role in setting up the Long Range Desert Group, a specially trained and equipped unit which made deep strikes into enemy territory in Libya. As Libya is lacking sandworms, this unit used trucks for long-range excursions, but like the Fremen, these troops tactically exploited the desert terrain and took special measures to conserve water. One of Bagnold's innovations in the build-up to war was the automotive equivalent of a Fremen stillsuit—the introduction of an external condenser for the radiators of their trucks. (In the 1920s and '30s it was normal for steam to be simply vented from an overheating radiator. The condenser allowed this vapor to be recovered and has since become a feature of all auto radiators.) Another innovation was the use of a sun compass. Metal equipment in their trucks would distort the readings of a magnetic compass requiring a stop to take bearings away from the vehicles, but with training a sun compass could be used on the move.

Dunes are generally active structures which are in motion. As documented by Bagnold in field observations and with wind-tunnel experiments, the dune moves forward via the motion of individual sand grains in a jumping process called saltation. When the wind speed is high enough and if the sand is dry, sand grains jump forward in a thin layer near the ground. As each grain flies forward in the wind and hits the ground, it kicks other grains, quickly building up a self-sustaining layer where the sand moves forward in the wind. Note that sand (grains above about a tenth of a millimeter across) is rarely lifted up into the air altogether—although this can happen with smaller particles in dust devils and dust storms. The way that this sand movement is concentrated near the ground is sometimes seen in the erosion of rocks or fenceposts in desert areas. They often are intact only a few dozen centimeters up, but are eaten away by sand abrasion just at ground level.

A dune may initially form where some small irregularity in the ground traps some of the migrating sand. Once a dune forms it persists and often grows. When the sand transport is in a single direction, the upwind side of the dune (the stoss slope) tends to be shallow and well-packed, often with ripples. The downwind side (the lee side or slipface) is continually avalanching forward and thus is loose. Desert warriors, whether Fremen or British, quickly learned that it was vastly more efficient to traverse a dunefield either along the stoss slopes or in areas between the dunes themselves. Movement across the slipface is highly fatiguing. As a Fremen remarked to Thufir Hawat, "We watched you come across the sand last night. You keep your force on the slipface of the dunes. Bad" (*Dune* 211).

The beautiful repeating patterns in dune fields result from the mutual interaction of the sand and wind. A dune alters the windflow downstream of itself, while the disturbed wind controls the sand motion which modifies the shape of a dune and pushes it forward. Dunes are therefore emergent structures, and it is possible to simulate their formation in quite simple computer programs. The complexity of natural dune patterns arises from the variability in sand supply and wind direction. Dune fields can show a variety of forms within a short distance.

There are many different types of dune. Perhaps the most familiar, at least by analogy with small scale ripples, is the transverse dune. This is a long dune that forms when there is ample sand and the wind direction is more or less constant. Where winds are constant, but the sand supply is weaker, isolated crescent-shaped dunes called *barchans* form. An intermediate type is *paraboloid*. If the wind direction is highly variable, an irregular star dune forms. Star dunes can be very large, up to 300 meters tall. Some examples are the dunes in Death Valley National Park in California. If, as often happens due to seasonal changes like monsoonal flow patterns, the wind comes from two predominant directions, a linear or longitudinal dune forms, aligned along the direction of net sand transport. This is roughly the average direction of the strongest winds.

Certain linear sand dunes, especially those with a sinuous form, are named *seif*, after the Arabic word for "sword." A large region completely covered in sand is called an *Erg*. Large, widely spaced megadunes (often with smaller dunes superposed on them) are called *draa*. Given the faintly Arabic etymology of many Fremen terms, it seems likely that these terms might be used on Arrakis. Another arid-lands feature, the salt pan, often goes by the Spanish term *playa*, or sometimes (especially near the coast) the Arabic term *sabka*.

In our Solar System, there are three principal worlds with extensive dunes—Earth, Mars, and Saturn's moon, Titan. (Venus has only a handful of detected duneforms, probably because there is very little sand.) Dunes cover 5 percent of the land surface of Earth but not uniformly. There are small and generally isolated dunes in many locations, notably near the edges of ice caps or former ice sheets, and of course coastal dunes where the action of the sea makes sand available. But most dunes are found in deserts at mid-latitudes, most particularly the Sahara and Namib (the continent of Africa is about 30 percent covered by dunes), as well as the Arabian and Australian deserts.

These terrestrial deserts generally appear in two belts at about twenty to 30 degrees north and south latitude. The reason is that the combination of the distribution of solar heating and the Earth's rotation lead to an atmospheric circulation pattern with a generally dry downwelling flow at these latitudes. Strong heating at the equator leads to upwelling of moist air, forming large thunderstorms. The storms precipitate out the water from the air as it rises, and thus the down-welling branch of this circulation (the Hadley cell) is dry. If Earth were to rotate more slowly, the down-welling would occur further from the equator, displacing the deserts poleward.

Field measurements show that the fastest-moving dunes (specifically, small barchans) can move at tens of meters per year. Dune migration can be disruptive to infrastructure such as roads, oil installations, and railways. Most notably, Nouakchott, the capital of Mauritania in northwest Africa, is threatened by massive dunes marching into its eastern margins. Experiments have been made with fabric fences to guide the wind. As a crisis measure, judicious placement of such structures can break up large dunes over a few weeks. However, since such fences would need to be replaced frequently, this is not feasible on a large scale. The use of plants to prevent the sand transport is the only practical approach, and even then relies on their being at least some moisture. In fact, Herbert's inspiration for writing *Dune* largely derived from encountering a U.S. Department of Agriculture experiment to use grasses to stabilize dunes in Oregon in 1957. In a letter to his agent, Herbert noted that dunes can "swallow whole cities, lakes, rivers, highways." Indeed, Herbert dedicated the book to dry land ecologists.

Contrary to the popular conception of the planet Mars, while it is indeed very dry, it is hardly covered in dunes. Patches of small dunes can be found in many places (notably in valleys and on crater floors), but extensive fields of dunes large enough to be observed easily from orbit are rather rare. A recent comprehensive survey documents the Martian dunefields (http://www.mars-dunes.org/).

The largest concentration of dunes on Mars is the Northern Polar Erg, a belt of dunes that forms a dark crescent surrounding the North Polar Cap. This sand sea has an area of some 700,000 square kilometers, about six times larger than the area of all other dunefields on Mars put together. In total, large dunefields cover less than 1 percent of Mars's surface area.

Martian dunes are predominantly transverse or barchan dunes. Wind patterns in the thin Martian air are fairly constant, often being controlled substantially by topography: cold air at night howls down canyons, for example. Note that faster wind speeds are needed to move sand on Mars. Even though the gravity on Mars is less than Earth, the air is 100 times less dense and so it needs to move ten times faster to yield the same force. Imagery from satellites orbiting Mars has yet to detect substantial movement of any Mars dunes, leading to speculation that the dunes formed in an earlier climate epoch with a thicker atmosphere or stronger winds, or that the dunes are somehow cemented in place by ice or salt.

Recent measurements by spacecraft have confirmed the suspicion that Mars has lots of hidden water frozen in large permafrost deposits and icecaps at the poles. There are many old river channels that suggest that water has catastrophically burst out from the ground and carved these channels at least occasionally. The transient seas formed by this flooding would have frozen over with the ice slowly evaporating to be re-deposited on the polar caps. Further evidence to support these ideas is measurements from orbiters and the NASA Mars exploration rovers that have identified minerals such as jarosite on the surface that are only formed in wet conditions. In this sense, the exploration of Mars parallels Dune: the evaporite minerals confirm a past more equable climate, just as Pardot Kynes found "a glaring white surprise in the open desert.... Salt. Now he was certain. There'd been open water on Arrakis—once."

Surprisingly, dunes have been found in abundance on Saturn's moon Titan (which is 5,150 kilometers in diameter, between Mercury and Mars in size). Titan is unique among the moons in the Solar System in that it has a thick atmosphere. It was speculated that its atmosphere, made mostly of nitrogen like Earth's, could easily move granular materials in its low gravity one-seventh of Earth's atmosphere, much like the Earth's moon's gravity. However, being so far from the sun (ten times further away than is Earth), winds on Titan would be very slow since sunlight is too faint to power strong winds.

A radar instrument on the NASA-European Cassini spacecraft observed vast fields of dunes on Titan in 2005. It seems that these dunes cover some 20 percent of Titan's surface, forming an irregular belt of dark sand seas between 30 degrees north and south. Among these large dark areas, some of which could be seen from Earth with large ten-meter-class telescopes like Keck, or with the Hubble Space Telescope, are Belet, Shangri-La, Senkyo, and Atzlan.

It turns out that Titan's near-surface winds may be higher than had been expected from the weak sunlight, because the effects of Saturn's gravity cause tidal currents in the atmosphere just as tides cause currents in the Earth's oceans. These tidal winds oscillate between two predominant directions, leading to the formation of longitudinal or linear dunes. The dunes on Titan are tens to hundreds of kilometers long, a kilometer or two wide, and often over 100 meters high, making them identical in size and shape to the linear dunes found, for example, in the Namib desert on Earth.

Interestingly, Titan's polar regions are dotted with lakes and even seas of liquid hydrocarbons. Methane and ethane (natural gas) are liquid at Titan's surface temperature of 94 degrees K. Scientists do not yet know what the sand is or how it is made, but suspicion is building that perhaps the sand is made from small organic haze particles that have stuck together in the lakes and then blown to lower latitudes when the lakes dry up. Another possibility is that very rare but possibly violent rainstorms (with liquid methane rain!) form river channels, carving out icy bedrock to form sandy material.

Titan's climate is only just beginning to be understood. The sand dunes are only found in a belt near the equator. It seems that on slowly rotating Titan, the Hadley circulation pattern tends to dry the ground at low latitudes, allowing the sand to move only there—like Arrakis, but over a narrower latitude range.

Whether a real planet could have sand dunes covering its surface out to 60 or 70 degrees from the equator like Arrakis is an interesting question. It might be that some set of planetary parameters such as size, rotation rate, equatorial tilt, and average temperature can lead to such a desiccation distribution, although whether these parameters are compatible with the handful of planetological details given of Arrakis in the books is another issue. The state-of-the-art of planetary climate simulation is now such that it should be possible to explore whether a planet with all the described features of Arrakis is really possible, and if so, for how long. No planet is static; stars evolve in luminosity, atmospheres lose light gases to space, carbon dioxide is absorbed by rocks, and the supply of geothermal heat from the planetary interior that drives volcanism and plate tectonics slowly declines. Just as Earth and Mars were very different places 3 billion years ago from what they are now, so Arrakis in the distant past and distant future will not be the same as was described in *Dune*.

Planetary evolution and rotation rate also play into the properties of the planet's interior. The Earth's strong magnetic dipole is generated by circulation of molten iron in its core acting as a dynamo, yet Venus, although similar in size and mass, has none. Arrakis apparently also has no dominant planetary-scale magnetic field, but instead Fremen can use a paracompass to exploit previously mapped magnetic anomalies to navigate on regional scales. Mars, incidentally, would be an ideal place to use a paracompass. It has no dynamo field (although it may have had one in the deep past, perhaps when it had water on its surface, too), but does have large stripes of magnetic rock in its southern highlands.

A vexing question in the evolution of a planetary landscape is origin of the vast deposits of sand. The very fact that giant worms could hide in Arrakis's sand places a lower limit on the sand thickness—many tens to hundreds of meters. This is comparable with some of the thickest large sand deposits on Earth. Sand results from the breakdown of bedrock. On a terrestrial planet this can occur by wave action on cliff-shores, through the action of rivers or glaciers. However, sand production is self-limiting—eventually there is enough sand to protect the bedrock from further erosion, so covering so much of the surface of Arrakis with deep sand would require some impressive sand transport to segregate areas of sand production from areas of deposition. As a counterpoint, Venus has a thick atmosphere which prevents its surface from day-night temperature changes and it lacks rivers, rain, and seas. Thus it appears to not have very much sand and therefore few dunes.

It is a curious feature of dry sand that it can be relatively transparent to radio waves (unless heavily loaded with iron minerals). Radar imaging from Earth orbit can reveal shallow buried structures or riverbeds beneath sand sheets, and deep-sounding radars are now in use on two Mars orbiters to probe the structure of the polar caps and detect geological features which are buried. One wonders why spice miners on Arrakis might not use some kind of ground-penetrating radar on their ornithopters instead of relying on observing wormsign.

The investigation of Mars's subsurface is emerging as a theme for future exploration. Engineers are confronting soil mechanics challenges that have apparently been solved by nature but were not explained in *Dune*. To attain locomotion through sand at depth requires a considerable expenditure of energy, one that cannot be realistically accounted for with what little is understood about the worm's diet. The simple experiment of trying to push a stick through even a meter of sand will illustrate the challenge. However, one must caution that nature can often find a way. Simplistic aerodynamic considerations cannot explain how the bumblebee can fly, yet clearly it can. (The bee exploits unsteady aerodynamic flow. Its wingbeats occur so quickly that the steady conditions for which airplanes are designed do not apply.) Perhaps worm locomotion can give us similar lessons about terramechanics. Indeed, a "self-hammering drill" that moves forward through the soil like an inchworm (although actually the device was referred to as a "Mole") was sent to Mars on the ill-fated European Beagle-2 mission.

It might be that the sand on Arrakis is modified on a large scale by the activity of worms. In fact, on Earth the structure and nutrient content of topsoil is significantly affected by earthworms. Fertile soil can support a population of up to about ten worms per square meter, each of whom can process (i.e., eat and excrete!) some four kilograms of soil per year. Put another way, worms can process about a meter's worth of soil depth in 500 years. Even in the unlikely case that Arrakis's worms had only the same small level of activity as earthworms, a 100-meter depth of sand would be processed in approximately 50,000 years, a planetologically short period.

Sand exhibits fascinating flow properties. If compacted, the grains lock together and the sand is quite rigid. Yet it can flow like water if disturbed by sufficiently strong vibration. In the absence of such "acoustic fluidization," the tidal sands mentioned by Herbert seem unlikely (and it is not obvious that the stars or moons near Arrakis are large or close enough to cause appreciable tides in any case, in water or sand). However, a real world counterpart to the "drum sands" exists: the so-called "singing sand" or "booming dunes" documented by Bagnold.

Exploration by spacecraft has shown that dunes occur on several planetary bodies. The point of this essay is to show that planets are beginning to be understood to a level where useful and instructive comparisons can be made between them, whether they are real or fictional. In addition to our fellow worlds around the sun, we now know of more than 100 planets around other stars. Scientists have even been inspired (Zahnle et al., 2005) to consider whether generally dry planets might have isolated near-polar regions which might be livable like Arrakis, even though the planet as a whole might not be.

In many ways, the landscape and climate of Arrakis resembles a warm version of present-day Mars. This is remarkable, given how little was known about Mars when Herbert wrote *Dune*. Telescopes showed that Mars has little polar caps, but Mariner 4, the first spacecraft encounter with Mars, occurred only in 1965, the year that *Dune* was published, and it is only in the last decade that evidence of Mars's hidden water has been found. While Mars is the obvious comparison, it is striking that the extent and thickness of Arrakis's massive sand deposits are more reminiscent of the sand seas of Titan. That world's dark sands have more than a whiff of spice in them, apparently being formed from complex organic molecules drizzling down from the sky. As is so often the case, truth is stranger than fiction.

RALPH D. LORENZ, Ph.D., is a native of Scotland, but resides in Columbia, MD. He has a B.Eng. in aerospace systems engineering from the University of Southampton, UK, and a Ph.D. in physics from the University of Kent, UK. He worked for the European Space Agency in the Netherlands on the design of the Huygens probe to Titan, and for over fifteen years has been involved in Mars flight projects and the Cassini spacecraft. He is a Fellow of the Royal Aeronautical Society, the Royal Astronomical Society, and the British Interplanetary Society. He has written several books, including *Spinning Flight*, *Space Systems Failures*, and *Lifting Titan's Veil*.

References

Bagnold, R. *Physics of Wind-Blown Sand and Desert Dunes*. Methuen, 1941.

Bagnold, R. *Sand, Wind and War, Memoirs of a Desert Explorer*. University of Arizona Press, 1990.

Herbert, F. *Dune*, Chilton Books, 1965 (Page numbers given here are from the Berkley Medallion 1975 edition).

Lancaster, N. *The Geomorphology of Desert Dunes*, Routeledge, 1996.

Lorenz, R. D. et al. "The Sand Seas of Titan : Cassini RADAR observations of Longitudinal Dunes," *Science*, 312, 724–727, 2006.

McKee, E. D. (ed) "A Study of Global Sand Seas," Geological Survey Professional Paper 1052, U.S. Government Printing Office, Washington D.C., 1979.

Shaw, W. B. *Long Range Desert Group*. Greenhill Books, 2000.

Zahnle, K., N. H. Sleep, Y. Abe and A. Abe-Ouchi. "Dune Exploration: Mars Allegories," American Geophysical Union, Fall Meeting 2005 abstract #P41D-08.

FROM SILVER FOX TO KWISATZ HADERACH: THE POSSIBILITIES OF SELECTIVE BREEDING PROGRAMS

Carol Hart, Ph.D.

He was known by many names—Paul Atreides, Mahdi, Muad'Dib—but most importantly he was the Kwisatz Haderach. Take ninety generations of selective breeding, add training and a little bit of spice, and you get a being with the prescience of a Guild Navigator, the computational power of a Mentat, and the insight of a Bene Gesserit. Could selective breeding produce such a superhuman? Carol Hart, Ph.D., tackles that very issue.

YOU ARE A MUTT. So am I. If there was an American Kennel Club for humans, no family, whether royalty or commoner, could consistently meet the breed standards. Two first-rate border collies with strong pedigrees will reliably produce a litter of good border collies: energetic, obedient (biddable) shepherds with a keenly intelligent gaze. In contrast, human parents-to-be may hope their newborn inherits her musical talents and not his nose, but they know very well that they have no control over the matter. We are mutts because we carry in our genes a motley assemblage of adaptive and maladaptive traits—an unselected ragtag of get-ahead and get-by genes. Undesirable traits are rapidly eliminated in wild species by natural selection and in domestic species by artificial selection, but we humans are under no such constraints. In our safe and sanitized civilization, we all muddle along with whatever genetic baggage we carry and pass it randomly to our children.

It is arguably myopic of us to care so much about the pedigrees of our pets and so little about our own. As readers of Frank Herbert will recognize, it is a theme running through the Dune novels, which implicitly ask how humans will continue to evolve in the absence of selection pressures.

In *Dune*, the theme is not immediately obvious because our attention is focused on a small elite group with remarkably advanced powers: the Mentat Hawat, the soldier-tactician Duncan, the Bene Gesserit–trained Jessica, and of course, her son, Paul, the Kwisatz Haderach. We get one brief glimpse of ordinary humans early in the book when a shuttle arrives bringing Duke Leto's men to Arrakis. "They carried their spacebags over their shoulders, shouting and roistering like students returning from vacation" (*Dune* 79). A few vivid fragments of dialogue—"Did you get a good look at this hole on the way down?" "Haven't you heard, stupid? No showers down here. You scrub your ass with sand!" (*Dune* 79)—tell us all we need to know. Some tens of thousands of years into the future, ordinary people are just like us. Depending on the reader's point of view, that similarity may be either reassuring or disappointing. In either case, it does make perfect sense. Without selection pressures of some kind, how would we change?

Natural Selection—The Desert and the Gom Jabbar

In *Dune*, the native Freman of Arrakis were admirable (and are truly free men) because they were under the tremendous selection pressure exerted by the hostile environment of desert and worm. In *God Emperor of Dune*, the greening of Arrakis had the unintended consequence of destroying the culture of the Freman, who survived only as museum exhibits and reservation Indians. Liberated from the continual struggle with the desert, they went slack and water-fat. Like army camouflage on clueless teenagers, their stillsuits were now shoddy imitations worn as fashion statements. In exchange for personal comfort and security, they relinquished a culture that (though violent by our standards) was superbly well adapted to their environment. Whatever our backgrounds, Asian, Native American, African, or European, our ancestors all made a similar transition from danger to security, from wild to tame. Like the citified Fremen, we are comfortable and safe. But can we really consider ourselves to be free, or even fully human, when we have lost the ability to survive on our own?

Again and again in the Dune novels, confronting a threat to survival is a crucial step to developing one's latent powers as human or superhuman. The elaborate training program of the Bene Gesserit sisterhood included two live-or-die ordeals: the agony box that all young novitiates must endure to test whether they are human or animal, and the spice agony that would endow them (if they survived) with the superhuman prescience of a Reverend Mother.

"Why do you test for human?" Paul asked Reverend Mother Mohiam after surviving her test. "To set you free," she answered (*Dune* 11). The ordeal by gom jabbar and the agony box began Paul's transformation: "Paul felt that he had been infected with terrible purpose. He did not know yet what the terrible purpose was" (*Dune* 11). Struggling to understand his experience, he asked Mohiam another question, then half-sarcastically answered it himself: "You say maybe I'm the…Kwisatz Haderach. What's that, a human gom jabbar?" (*Dune* 13).

Mohiam told Paul that the Kwisatz Haderach would have complete prescience to peer into all the avenues of the *past*, male and female. As they would discover to their shock, the Bene Gesserit had unconsciously bred Paul for abilities they did not suspect. The Kwisatz Haderach would be able to peer into all the avenues of the *future* and to be, in fact, a human gom jabbar, the goad that forces a choice between evolving as humans or perishing as animals.

On Arrakis, after his mother and he fled into the desert to escape the Baron Harkonnen and the Emperor's Sardaukar, Paul's powers of prescience expanded with each successive threat to his survival. He saw yet more clearly his "terrible purpose" (*Dune* 199), a path that would bring horrible destruction to humankind, and shrank from fulfilling it. It was a destiny that Leto II, his son, embraced as his "golden path," to be a predator, to give humanity "a lesson their bones will remember" (*God Emperor of Dune* 185) in order to rouse them from their tameness. The "terrible purpose" and the "golden path" are a renewal of selection pressures to force humanity to scatter, to colonize new worlds, and to evolve new forms and powers.

Artificial Selection—
Where the Tame Things Are

The elite sisterhood of the Bene Gesserit had their own, gentler answer to the problem of human evolution: the selective breeding, over some ninety generations, of a fully prescient human who would be the Kwisatz Haderach. In the process, they discovered two things that breeders here on Old Earth have also found: first, that seemingly complex traits can be specified by single genes, and second, that selecting for one desirable trait may bring a host of other unforeseen traits with it. The Kwisatz Haderach came a generation earlier than planned and ultimately, as a human gom jabbar, destroyed the power of the sisterhood.

Because we see ourselves as complex, we assume special abilities must be polygenic, the product of many distinct genes acting together. We may be wrong about that. Some of the seemingly spectacular differences in domesticated breeds may be the product of just one or two genes. Think of the hundred or more dog breeds on parade annually at the Westminster Kennel Club Dog Show—Pekingese to pointers to poodles, setters and schnauzers and shepherds. A recent study appearing in *Science* found that the enormous size variations in domestic dog breeds arise from variations (polymorphisms) in one growth factor gene. Only a few nucleotides, a tiny snippet inside a single gene, specify the dimensions for a Chihuahua versus a Great Dane. We think human talents are complex traits because we are mutts. If math prodigies only mated with math prodigies, we might arrive at a clan of Mentats in just a few generations. Or if acrobats were only sexually attracted to other acrobats, they might soon possess the phenomenal reflexes (but hopefully not the aggression) of those human Rottweilers, the Honored Matres who attempted to hunt the Bene Gesserit and the Bene Tleilax into extinction in *Chapterhouse: Dune*.

Except for us, most of the plants and animals in our immediate environment, and certainly almost all of our foods, are the products of artificial selection. Yet the history of our selective breeding efforts is mostly a story of lucky accidents and unintended consequences. The jungle fowl *Gallus gallus* was originally domesticated for the oldest of all spectator sports, cockfighting. Selecting cocks for their ferocity led to the spin-off benefit of a hen who lays nonfertile eggs on a near-daily basis (*Gallus domesticus*). Less luckily, recent short-sighted commercial efforts to develop leaner hogs that could be marketed as "the other white meat" resulted in anxiety-prone animals that drop dead when badly startled.

Our first and finest achievement of domestication is, of course, the dog, who has tagged along at our side for some 15,000 years. It is still debated whether humans domesticated wolves or wolves domesticated themselves by moving into the rich ecological niche of scavenging human settlements. In either case, the evolution from wolf to dog, from dangerous predator to best friend, seems extraordinary. Yet we now know it can happen in just a few generations if the selection pressure is sufficiently intense.

In the 1960s, researchers at the Russian Institute of Cytology and Genetics began an ambitious program to study the genetics of domestication by intensively breeding silver foxes for tameness. The single breeding criterion was a behavioral tolerance for humans, but it was strictly applied. At each successive generation, only a small minority, 3 percent of males and less than 10 percent of females, was allowed to breed. In the sixth generation the first tame foxes appeared, just 2 percent of the population.

Over the course of just forty-odd generations, the Russian researchers have produced a breed of fox that strikingly mimics the family dog. Selecting for the apparently simple trait of tameness had remarkable effects upon the animal's overall appearance and behavior. The tame foxes bark and whine for attention, wag their big fluffy fox tails in excitement, and roll over to get their tummies petted. Physically they are often dog-like as well, with floppy ears, piebald markings (much like a border collie), a shorter, curled tail, and a puppyish rather than foxy face.

The behavioral changes are quite complex. One recent study found that the tame foxes share the canine ability to correctly interpret human gestures and body language (something the bigger-brained apes do only poorly). The researchers speculated that breeding the fox for a lack of fear might have permitted their latent abilities to observe to become more fully developed. The Bene Gesseret similarly saw the importance of quelling fear to enable observation.

Ongoing studies of the domesticated silver fox are focused on understanding the genetics of domestication in general, which might include the domestication of humans as well as dogs. However slowly and haphazardly, we have been selecting ourselves for docility. Despite crime rates that oscillate up and down, over the course of many millennia have been gaining in the capacity to cooperate and to control aggressive impulses, a theme recently developed by Steven Pinker in an essay for *The New Republic*. After all, it is no longer socially acceptable to brain an annoying rival with a club, or to stab a sibling who reaches for the last cookie.

The Bene Gesserit were within a generation of completing their program. However, Jessica did not do their bidding, and her son and grandson were yet more rebellious. Arguably, the Bene Gesserit breeding program failed because the Reverend Mothers paid too little attention to preserving the domestication trait. They recognized too late the dangers of breeding a superhuman who, unlike the well-bred dog, is not biddable. Perhaps we ought to settle for being mutts after all.

CAROL HART, Ph.D., is a freelance health and science writer based in Narberth, PA, just outside of Philadelphia. She is the author of *Good Food Tastes Good: An Argument for Trusting Your Senses and Ignoring the Nutritionists* (forthcoming, SpringStreet Books) and *Secrets of Serotonin* (St. Martin's Press, 1996, with a revised and expanded second edition forthcoming in early 2008).

References

Hare, B., I. Plyusnina, N. Ignacio et al. "Social Cognitive Evolution in Captive Foxes Is a Correlated Byproduct of Experimental Domestication." *Current Biology* 2005; 15:226–230.

Herbert, F. *Dune*. New York: Berkley Books, 1977.

————. *God Emperor of Dune*. New York: Berkley Books, 1981.

Lowry, C. A., J. H. Hollis, A. de Vries et al. "Identification of an Immune-Responsive Mesolimbocortical Serotonergic System: Potential Role in Regulation of Emotional Behavior." *Neuroscience* 2007.

Pinker, S. "A History of Violence." *The New Republic*, March 19, 2007 (accessed online).

Popova, N. K., N. N. Voitenko, A.V. Kulikov, D.F. Avgustinovich. "Evidence for the Involvement of Central Serotonin in Mechanism of Domestication of Silver Foxes." *Pharmacology, Biochemistry and Behavior* 1991; 40:751–756.

Sutter, NB, C.D. Bustamante, K. Chase et al. "A Single IGF1 Allele Is a Major Determinant of Small Size in Dogs." *Science* 2007; 316:112–115.

Trut, L.N., I.Z. Plyusnina, I.N. Oskina. "An Experiment on Fox Domestication and Debatable Issues of Evolution of the Dog." *Russian Journal of Genetics* 2004; 40:644–655.

Trut, L. "Early Canid Domestication: The Farm Fox Experiment." *American Scientist* 1999; 87:160–169.

EVOLUTION BY ANY MEANS ON DUNE

Sandy Field, Ph.D.

In his 1859 book On the Origin of Species, *Charles Darwin put forth the idea that species pass on desirable traits to their descendents, and evolve, by means of natural selection. What if human beings are manipulating the process by selective breeding and gene manipulation. Can we still call this evolution? Sandy Field, Ph.D., discusses unnatural selection in the Duniverse.*

Natural selection has been described as an environment selectively screening for those who will have progeny. Where humans are concerned, though, this is an extremely limiting viewpoint. Reproduction by sex tends toward experiment and innovation. It raises many questions, including the ancient one about whether environment is a selective agent after the variation occurs, or whether environment plays a pre-selective role in determining the variations which it screens. Dune did not really answer those questions; it merely raised new questions which Leto and the Sisterhood may attempt to answer over the next five hundred generations.

—The Dune Catastrophe, After Harq al-Ada, *Children of Dune*

GENETIC MANIPULATION is a powerful theme throughout the Dune series. From the Bene Gesserit desire to create the Kwisatz haderach through carefully planned breeding, to the Tleilaxu and their unusual methods of reproduction, it is clear that Frank Herbert viewed the future as a place where science and medicine would create new life forms outside the realm of typical natural selection. In Dune, Herbert created a universe where "desired" traits could be enhanced through many carefully controlled generations or created through mechanical manipulation of genetic molecules. New abilities could also be "naturally" selected by subjecting humans to extreme environmental pressures. His vision includes some ideas based firmly in scientific reality and others that are science fictional flights of fancy. When the first Dune book was written in 1965, genetics and the role of deoxyribonucleic acid (DNA) in the genetic basis of heredity were the hot topics of the day. Darwin's evolutionary theories, published in 1859, had borne up well under scrutiny and now formed a basic tenet of many scientific fields from biology to geology. The structure of DNA had been elucidated by Watson and Crick in 1953 and provided an elegant explanation for how genes could be passed from parent to offspring to provide the blueprint for life. The processes by which new species developed and adaptations emerged over time were now clearly understood to be a result of random changes in DNA followed by natural selection in the environment.

The background for understanding genetic processes at the molecular level was also well established by the time of the first Dune book in 1965 and even more clearly by the time Herbert began to explore alternate forms of reproduction with the Tleilaxu in *Heretics of Dune*, which was published in 1984. Research had shown that the information contained in the four paired bases of DNA provides a template from which to build proteins, the molecules that run our every biological process. The sequence of the nucleic acids in DNA can be read like a book with only three-letter words. Each triplet of bases represents a three-letter code for each amino acid that is to be added to a growing chain. Joined together, these amino acids build enzymes and structural proteins that are the machinery of life.

Although many non-scientists seize on the phrase "survival of the fittest" and think of evolution as the battle between individuals to survive in their environment and to reproduce at any cost, the introductory quote above from *Children of Dune* shows that Frank Herbert clearly understood that creation of the "fittest" is predetermined by one's genes as part of the random mutation and recombination of DNA during DNA replication, meiosis, and sexual reproduction. Natural selection of the resulting genetic attributes by the environment occurs after the organism has already been "adapted."

Frank Herbert uses both of these basic tools of evolution, random mutation, and natural selection in his creation of new races for the worlds of Dune. The Bene Gesserit and Tleilaxu make use of their knowledge of how to manipulate genes through breeding and cloning techniques to create specific traits in their offspring. The Padishah Emperor enhances his selective ability by using oppression and harsh conditions to generate competition on Salusa Secundus and thus creates a gene pool from which to choose his army of Sardaukar fighters. The Fremen are also naturally selected by the brutal adverse environment on Dune, further showing how survival genes can be ruthlessly selected under the most stringent conditions.

During the time the Dune series was being written, the true modern genomic revolution was still yet to come, with our more recent understanding of organisms on a whole genome basis and even deeper understanding of the relatedness of all life on the planet from microbe to multicellular organism. In addition, understanding of gene regulation, the influence of gene sequence on behavior, and the complex interplay of genetics and environment are still under investigation. However, Frank Herbert, based on sound basic genetics and some extrapolation, provides a believable and fantastic world in his Dune books.

The Bene Gesserit Breeding Program

Over the course of many millennia the sisters of the Bene Gesserit worked tirelessly to breed certain traits into their genetic lines of humans in order to amplify the traits they deemed desirable and to discard those they found to be unworthy. They were particularly interested in the Atreides family genes and had worked for many hundreds of years to create new "strains" of superior humans that would have special characteristics while remaining submissive to the Bene Gesserit plan. Creation of super beings is a common theme in science fiction, and it usually ends in disaster. Let's take a look at how it worked for the Sisterhood.

Genetic crossing of good breeding stock has been practiced here on Earth for many hundreds of years with food crops, animal stocks, and to some extent with humans. In general, tinkering with genetic traits has been a successful practice for people. For example, we have selected for bigger, hardier, more insect-resistant fruit with enhanced fleshy parts and we have selected for chickens with increasingly bigger breasts. So far, so good.

However, extensive breeding of dogs has revealed that there is a cost for this type of selective breeding. Some dog breeds have begun to exhibit an array of inherited diseases that have become typical of that breed. For example, German Shepherds tend to have problems with paralysis in their back legs, blood clotting problems (hemophilia), and heart disease, while Dalmatians inherit nerve and kidney problems.

Why is this? It is due to the buildup of recessive alleles found in the population and usually retained only at a low level. Organisms that reproduce by sexual reproduction generally have two copies of each chromosome that contain one copy of each gene called an allele. Alleles can be identical within an individual but more often are different from one another because one copy comes from each parent. One copy may be "dominant" in that the trait is always expressed if the allele is present, while the other copy may be "recessive" in which the trait is only expressed if a dominant allele is not present. Most of us are familiar with this in terms of eye color, where brown eyes are dominant over blue eyes.

Genetically, these alternate gene forms serve to provide a template for diversity of species that survive by constantly changing. This change is the random re-assortment that occurs during sexual reproduction and creates the ability to be "naturally selected" should those changes end with a positive result. However, if both parents are too closely related and share the same recessive alleles at a particular gene location, there is no diversity created and the potentially harmful gene will be passed on to all offspring rather than only one-fourth of the offspring as is usually the case with a recessive gene.

Recessive genes can be a problem with selective breeding of humans as well. In most cases, selection of specific breeding partners and limiting of genetic lines, as is the case in royal families and cases of religious separatism, have only served to amplify negative traits. This is what happened with the hemophilia genes in the royal families of Europe. The harmful mutation for this blood clotting disorder likely originated with Queen Victoria of England in the 1800s. Once generated, however, the gene was spread through all of the royal families in Europe due to a selective breeding situation created by intermarriage only between royals. This resulted in generations of hemophiliacs in all of the royal families in Europe and is still present in some of them today.

The same problem has occurred in some religious groups existent today in the United States. Restriction of breeding partners through strict religious requirements has concentrated genes in the old-order Amish population that afflict their carriers with various disorders, including hemophilia, anemia, immune system disorders, albinism, and dwarfism. Similar afflictions, including developmental problems, kidney disorders, and the concentration of other rare diseases have also occurred in contemporary Mormon populations due to nonrandom mating habits such as inbreeding and polygamy.

So how did it work out for the Bene Gesserit? Presumably, with all their careful planning and tracking, they could avoid the pitfalls associated with working in a limited gene pool. The goal of their program was the development of a male Bene Gesserit equivalent, the Kwisatz Haderach. They had planned to breed an Atreides daughter with a Harkonnen son, Feyd-Rautha, to create the desired combination, but Lady Jessica's determination to give her beloved Duke Leto Atreides a son instead of the daughter the Sisterhood had ordered created a problem. They wanted to create a male that could survive the spice agony and have prescient abilities but who would also be one of them. Creation of Paul Atreides seemed to have achieved their goals, but he is definitely not one of them. Indeed, it seems shortsighted to create such a superior being and then be surprised when he doesn't do their bidding! Judging from how things turned out with Paul's children, it seems unlikely that the outcome would have been much different a generation later.

There are also other genetic questions that come out of this situation. First, how do the Bene Gesserit control what sex their baby will be? Sex in humans is determined by the father. This is because the father has two types of sex chromosomes, X and Y, and each sperm gets one of them. Mothers have two copies of the same sex chromosome, both X, and so they can only provide an X chromosome to the embryo. It is the addition of the father's X or Y that determines whether the baby will be a boy, XY, or a girl, XX. We must assume here that the Bene Gesserit can determine the chromosome type carried by each sperm and regulate their access to the egg accordingly.

Second, how could prescience develop as a genetic trait? This is never explained in the books and represents pure science fiction. The ability to see into the future has always been a favorite science fiction theme but there is no basis in scientific fact for this. Humans currently have no documented prescient abilities.

Inheriting the Memories of All of Your Ancestors

Another of the main tenets of the Dune series is the spice-induced awareness of one's ancestors in susceptible individuals. Historically, this had been the sole realm of the Bene Gesserit Reverend Mothers. However, with the ritualistic, spice-induced conversion of the pregnant Jessica to a Reverend Mother, her unborn child, Alia, becomes instantly aware of herself in the womb and aware of the memories of all of their collective ancestors, becoming the first of the Atreides "pre-born." This is followed by the births of Paul's children Ghanima and Leto II who are also "pre-born."

Up to this point, we have seen that Frank Herbert combined existing scientific information with some creative extrapolation for development of the people in the worlds of Dune. With the introduction of the concept of "other memories" and "pre-born" lives, however, he takes off into the unknown. Here's why.

Our understanding of how memories are generated and retrieved has been essentially the same since 1965. Brain researchers at that time determined that memories are generated by the creation of new connections between neurons within an organism through experience and, although the process may involve the actions of nucleic acids or proteins, the memories themselves are stored as a type of neuronal code. This code is based on the proximity and firing patterns of neurons in the brain that represent memories. Current understanding of memory has filled in some of the details of this process but the basics have remained unchanged. There is no evidence that memories are inherited by offspring.

However, there are two possible pathways which could be extrapolated to include the inheritance of memories. One is the "instinctive" memory system that we are all familiar with. Our muscles instinctively know how to perform certain actions, like suckling a breast, without being taught. The same is true for other animals. They have inborn "memories" of how to do certain things. However, this is thought to be the result of the way that the neurons are laid down during embryonic development rather than being what we would call true memory. As a developmental process, instinctive memory can be passed to offspring because genes determine developmental pathways. But this is not memory.

The second possible way to inherit memories could be through what is called cytoplasmic inheritance. This extrachromosomal (not inherited with the chromosomes) inheritance of various traits is now known to be associated with plasmid DNA in bacteria and with mitochondrial DNA in eukaryotic (nucleated) organisms. In 1965, although it was known that certain traits appeared to be inherited from sources other than the main chromosome, it was only suspected that mitochondrial DNA might be responsible. However, this concept can be extrapolated to include what the Bene Gesserit call "other memories."

Mitochondria are small organelles within cells that make energy for us. They are thought to be the result of an ancient event in which we developed a symbiotic relationship with a bacterium that took up residence within our cells. Over time, these bacteria have become a part of us, but because they are bacteria, they still contain DNA and they still have the ability to replicate themselves. Mitochondria divide in a different fashion from everything else in the cell and are passed down as part of the mother's contribution to the embryo in the cytoplasm of the egg, not with the chromosomes. Activation of mitochondrial DNA that you inherited from your mother and so on down the line could, theoretically, account for the passage of genes that cause the creation of new neuronal connections containing her memories.

The real problem with this concept is the storage of all these new neuronal connections. With a limited cranial volume, where would these extra spice-induced neurons be kept? Envisioning a system where one's existing neurons change to accommodate other memories is very complex. In addition, Bene Gesserit Reverend Mothers and the Tyrant, Leto II, are reported to sit thinking and running back through their other memories—this does not evoke a vision of a neuronal reorganization system. This is also true of Alia, who is eventually taken over by her ancestor the Baron Harkonnen, and is present within her psyche at the same time as the Baron and others. All of these people in the book report the internal clamor of past lives as well. This implies a system where other memories are always present, not activated as needed, and such a system that concurs with currently accepted science is difficult to imagine without a generation of new neurons and a place to store them.

The Tleilaxu—Face Dancers, Axlotl Tanks, Gholas, and Spice

"Dirty Tleilaxu!"

—Duncan Idaho ghola, *Heretics of Dune*

The Tleilaxu are introduced early in the Dune series with the character of Scytale and the complex plot to assassinate Paul Muad'Dib using the first Duncan Idaho ghola. Scytale is a Tleilaxu Face Dancer who can assume the appearance and voice of anyone he chooses. The Tleilaxu are the only Dune race with this ability and they use it to infiltrate other worlds by replacing people with their spies. They are also the only race that can generate gholas. Gholas are reconditioned bodies that are made from the cells of a dead person in the mysterious Tleilaxu Axlotl tanks. Later, they learn how to make spice in these tanks.

The Duncan Idaho ghola is recognized immediately as a potential assassin but is actually a Tleilaxu experiment. As a result of the extreme pressure put on the ghola by the instruction to kill Paul (who his Duncan Idaho self died to save), the ghola regains his memories as Duncan Idaho. The Tleilaxu immediately see that they will be able to repeat this awakening process with other gholas, although we don't find out until *Chapterhouse: Dune* that they themselves are actually gholas and have been maintaining themselves this way for the many generations since the first Duncan Idaho ghola.

The implications of all of the various Tleilaxu genetic manipulations are not fully explored until the fifth book, *Heretics of Dune*, where the Tleilaxu are featured more fully and the details of the Face Dancers, gholas, and Axlotl tanks are revealed.

Face Dancers

The Tleilaxu Face Dancers are capable of making radical changes in their appearance and voice to mimic those of another person. What would be required, genetically and biologically, to wholly change your form into that of another? The changes that Tleilaxu Face Dancers undergo are described as being physical, in terms of height, weight, voice, and facial features, and then later as mental, with a new generation of Face Dancers who can take over the memories of a dying person and become even more like the original.

Physical attributes involve the organized distribution of the body into three body layers—the cells and tissues lining our gut and other organs (endoderm), the skin and neuronal layer (ectoderm), and our structural components—skeleton and musculature (mesoderm). The organization of cell types is determined during development in a sequential process that is dictated by our genes. The regulated expression of genes that are specific both to the species and each individual determines the developmental pathways that cells take to become various parts of the body. Cellular fates inherited from your parents determine the specific structure of your nose, for example, and follow genetic instructions that likely include influence from both parents' genes.

Once these determinations have been made, cells progress from a pluripotent state, in which they can become any cell type, to what is termed differentiated. Differentiated cells are those which have already gone down the developmental pathway to become a certain cell type. Chemical signals that are received by the cell during the process of differentiation tell it where to go, what cell type to become, and whether to keep dividing or stop dividing. Once a cell has become differentiated, it cannot be returned to its original pluripotent state. For example, the differentiation pathway to become a skin cell differs from that of a muscle cell, and you can't make skin cells turn into muscle cells. However, you can take a pluripotent stem cell and make it become any cell type if you know the right signals. This is the reason why stem cell research to regenerate lost or damaged tissue requires embryonic stem cells—they still are pluripotent and can be manipulated to become any cell type.

So what have the Tleilaxu done with this knowledge? If we assume that they are bound by the inability to change the fate of a differentiated cell currently forming some specific tissue type, such as muscle, we can try to figure out how they might change their appearance using only the cell types at their disposal; those they were born with. Face dancing, then, would require the Tleilaxu to have evolved a way of moving or migrating their cells to new locations on their bodies—in ways they could control by thinking about where they wanted them to go.

The books never really address how long it takes for a Face Dancer to assume another identity but the attack on the Honored Matres no-ship in *Heretics of Dune* provides a good example. The Tleilaxu master, Waaf, has just killed the Honored Matre. His Face Dancer guards come in and assume her identity and those of her guards in a relatively short time. This implies that the Face Dancers can imprint the new face and make the changes relatively quickly, probably in a matter of minutes. This would require reorganization of skin cells as well as musculature and skeletal elements in a very short time. This type of reorganization would require the breaking of cell to cell bonds that hold tissues together and then movement of cell types around to new locations. Also, one would imagine that changes in body height and weight would require changes in bone structure, body fat, and water content. Voice changes could be the result of a very flexible voice box and do not really require any extraordinary abilities.

What kind of genetic change might allow this amazing ability to completely change one's outward appearance? Most of the genetic changes that have been studied provide for incremental differentiation, generation by generation, over time. Evolution of a fish into a land animal with legs and lungs, for example, would take many, many generations and adaptations over time. On a shorter time scale, as an organism develops from a newborn to an adult, body changes occur with various forms occurring over the course of one's life. These changes can be startling—the larval form of an organism may be completely different from the adult form, but these changes also occur over a period of time. They may take weeks or months to occur and are on a developmentally predetermined path. They are not voluntary as with the Face Dancers.

Some organisms undergo annual changes in body appearance. Both animals and plants can change their appearance in different seasons.

For example, some mammals can change the color of their fur during winter to adapt to the changed environment. Chameleons can quickly change the color of their body based on their environment, but this is not voluntary. Is there an example of an organism that can change its body appearance at will?

There are many single-celled animals that can change their cell shape quickly in response to environmental changes. Amoebae, for example, move by reorganizing their internal cytoskeleton, their cell membrane, and their cytoplasm in a concerted manner. Their bodies change shape and move across a surface in response to an external signal, such as a chemical found in food. Amoebae can abruptly change direction and move toward or away from that signal, depending on its source.

Extrapolating from this single-celled response, one can imagine a genetic change that might allow for human cells to retain their ability to move in such a manner in response to a chemical change. Determination of cell fates during development requires the movement of cells to various locations to complete the emerging organism, so we know the cells have the internal machinery to move. It is simply a matter of providing the right signal. The concerted action of newly created hormones selected genetically by the Tleilaxu over many generations could act to allow different cell types to move when prompted by neurological signals.

Hormone signaling is a complex system regulating everything from our internal body temperature to our desire to breed through the action of chemical messengers that travel in our blood. In general, the system is not voluntary and, in fact, we are more likely to be directed by our hormones than the other way around. However, there are instances when emotions can cause the release of hormones, such as adrenaline in the classic "fight or flight" response to danger. It is not difficult to imagine adaptation over many generations of a people who could learn to control some of their own hormones.

Face dancing, then, could be a genetically derived ability to generate specific hormones at will, which allow for the concerted movement of skin, muscle, bone, and other cells to new locations to create the appearance of another person. Specific hormonal signals would be regulated by neuronal processes generated by the Face Dancers' observation of the face they wanted to mimic. This is consistent with everything Herbert reveals about the Face Dancers, and he even goes so far as to explain the Bene Gesserit ability to detect Face Dancers through their ability to detect Face Dancer hormones.

The New Face Dancers

Now, what of the new Face Dancers who are introduced in *Heretics of Dune*? These new and improved versions have the ability to both change their appearance and take over the memories of another person, usually a person of power who has been killed in order to be replaced by a Tleilaxian spy. These new Face Dancers are also harder for the Bene Gesserit to detect, presumably due to refinement of the hormones regulating the process.

What does this new talent require? There are two aspects to this new ability. First, they must be able to reorganize their neurons to record the new memories, and second, they must have the ability to transfer them from a dead or dying body.

The first part, while somewhat far-fetched biologically, can be explained again by the concerted reorganization of neurons. Memories are thought to be created by the generation of new associations between neurons that record the information of the memory in terms of the way they associate with each other, the way they associate with others, and the timing of the firing of those neurons in a meaningful way that conveys information to the brain. The mental image created by this neuronal spatial and activation pattern is what we term a memory.

If we are comfortable with the idea that the Tleilaxu have already evolved the ability to move their skin and muscle cells around, it is only a small step to consider that they could have developed a new hormone to move their neurons around and "create" new memories. How the old memories are affected is also addressed by Herbert, who tells us in *Heretics of Dune* that the Face Dancer who takes over the High Priest on Rakis, Tuek, actually eventually becomes Tuek and forgets his previous Face Dancer identity. Maybe a little more genetic tinkering is required to perfect this new generation of Face Dancers.

The second ability, that of transferring the memories or determining the spatial and activational organization of another's brain while they are dead or dying, is harder to explain in terms of a newly evolved ability. This is presumably related to the Bene Gesserit and Atreides abilities to both transfer memories and retain ancestral memories with the help of the spice. However, the Face Dancers can do it without spice and don't get the ancestors or the prescient abilities. One could invoke the ability to detect and reproduce some complex chemical signal that the dying person creates but there is no scientific basis for this. Herbert does not attempt to explain this ability to any great extent and leaves the basis for this ability unknown.

Axlotl Tanks—Making Gholas, Making Spice

The Tleilaxu have also tinkered with their own genetics in ways that reveal themselves throughout *Heretics of Dune* and *Chapterhouse Dune*. As the Honored Matres's assault becomes focused on the Bene Gesserit and others who can resist their sexual temptations, the Bene Gesserit end up holding the last Bene Tleilaxu master, a Scytale ghola, as a captive on Chapterhouse. In their attempt to replace the spice source lost during the destruction of both Rakis and the Tleilaxians, the Bene Gesserit seek to recreate the ability to generate gholas and spice from Axlotl tanks. These are revealed in *Chapterhouse Dune* to be some combination of female Tleilaxian flesh and nutritional systems. The Bene Gesserit actually convince sisters to volunteer for this unsavory duty but they cannot get Scytale, their captive Tleilaxian, to provide the necessary information to achieve spice generation. However, they do get enough information from him to generate the Bashar Miles Teg and presumably other people who have died as well.

The scientific details of the Axlotl tanks are unclear and mysterious throughout the series. This is ostensibly due to the fact that the Tleilaxu consider their monopoly on the tanks to be their biggest source of power over others. This makes them very secretive about them. Their development of spice from the tanks just confirms their importance. However, some hints are provided regarding the tanks that make it seem likely that the Tleilaxians may not be entirely proud of the methods used to accomplish their genetic feats. In *Heretics of Dune*, when a Bene Gesserit has a discussion with Master Waaf regarding the failing of the tanks, the sisters remark that he behaves as if he is being asked to discuss a deformed child or a mad uncle. Later, images of female Tleilaxians (and then Bene Gesserit) hooked up to a series of tubes to provide nutritional fluids, and mention of environmental controls fill in some of the unpleasant details regarding the tanks. In *Heretics of Dune*, Duncan Idaho remembers "a great mound of female flesh—monstrous in her almost immobile grossness...a maze of dark tubes linked her body to giant metal containers."

We can surmise that the fertilization of the "tanks" must be via artificial means. In *God Emperor of Dune*, Tleilaxu reproduction is mentioned as being a "pattern of gene surgery followed by artificial insemination." Later, in *Heretics of Dune*, the Bene Gesserit observe that the Tleilaxian "sperm does not carry forward in straight genetic fashion." The Tleilaxu Face Dancers are mules and cannot reproduce themselves. The Tleilaxu generate all sorts of gholas from the tanks, from servants to reproductions of themselves. This suggests a great diversity of form which must be provided by genetic manipulation prior to implantation in the tank. It is never implied that the females who provide the incubation chambers contribute to the genetic components of the growing organism.

These techniques, while not actually in existence, are certainly easy to imagine in our current age of *in vitro* fertilization and cloning, the basics of which have been around for many years. Genes could be derived from dead tissue, or other organisms could be artificially implanted into an egg from a female; or perhaps even a completely artificial embryo that could then be sent down the path of development by signals within the genes or derived from the maze of dark tubes linked to the tanks. The Tleilaxu have presumably mastered all sorts of abilities to manipulate development. They are able to implant reactions that can be activated later in the gholas' life and they can create gholas that act as servants and gholas that are enhanced with various abilities. Other races find their methods distasteful, but that does not stop them from using Tleilaxu gholas when they have someone they want to regenerate.

Generation of spice is another matter. We are left at the end of *Chapterhouse Dune* with the Bene Gesserit's success in creating spice from sandworms on their new desert planet. Scytale, clinging to his idea of regenerating all the Tleilaxians he has stored in the nullentropy tube implanted in his skin, has not given away the secrets to making spice in the Axlotl tanks but there is mention in *Heretics of Dune* that "a human body produced the spice in the Axlotl tanks." Since the original spice comes from the complex biology of sandworms, it is reasonable to assume that the Tleilaxu have somehow figured out which genes are involved and have implanted them into Tleilaxu females who are serving as tanks. Maintaining the correct conditions and adding the required building blocks would presumably generate spice, but since spice seems to be a metabolic byproduct rather than a new organism, it is unclear why they need the tanks at all. If they understand the chemistry, why do they need the female flesh? Certainly, it makes for good science fiction, and the vague images we are given of the Axlotl tanks only serve to make them more scandalous and repellent.

In his Dune series, Frank Herbert provides examples of all aspects of the evolutionary process in his creation of races for his various worlds. Through the use of solid genetic principles and thoughtful extrapolation, Herbert creates a universe where evolution is progressing by all possible means. The tenuous and often confounding nature of human survival is carried throughout the books and is explored to its fullest. This leaves the reader wondering which race will prevail, with humans buffeted by violence, religion, and hardship and adapting new survival tactics along the way.

SANDY FIELD, Ph.D., is a freelance science writer based in Lewisburg, PA. She holds a BS degree in genetics from the University of California, Davis and a Ph.D. in biochemistry, cell and molecular biology from Cornell University in Ithaca, NY. She specializes in continuing medical education, writing in the areas of oncology and infectious disease. Her other passions include science fiction, forensic science, San Francisco Giants baseball, bicycling, Tae Kwon Do, and cooking for her family. Visit her Web site at http://www.fieldscientific.com.

References

Darwin, Charles. *On the Origin of Species by Means of Natural Selection, or the Preservation of Favoured Races in the Struggle for Life.* London: John Murray 1st edition, 1859.

Watson, JD and FH Crick. "Genetical implications of the structure of deoxyribonucleic acid": *Nature*, 1953 May 30; 171(4361): 964–7.

Inherited Diseases in Dogs Database: http://server.vet.cam.ac.uk/index.html

THE ANTHROPOLOGY OF *DUNE*

Sharlotte Neely, Ph.D.

(All citations are from *The Illustrated Dune* by Frank Herbert.)

Anthropology has been described as "the most scientific of the humanities, and the most humanistic of the sciences." Sharlotte Neely, Ph.D., discusses what it means to be human within the Duniverse, and unlike the Bene Gesserit's narrow definition, the focus here has little to do with mere pain tolerance.

ANTHROPOLOGY'S APPEAL, unlike sociology and other disciplines that study people, lies in its breadth. Frank Herbert realized this and made use of the fact in *Dune*. If a topic somehow relates to people, it is anthropology. In other words, anthropology is incredibly holistic. One can be an anthropologist and study humans as either social or physical beings or both, anywhere on the planet, during any time period. The discipline is so broad that even those who study wild chimpanzees, humans' closest living relatives, are anthropologists.

Cultural anthropologists study the behaviors and traditions of societies all over the planet, concentrating on the non-Western world so often overlooked by other disciplines. Herbert fulfilled this focus of cultural anthropology with the creation of societies as different as the one ruled by the kindly Duke Leto Atreides on Caladan and the one ruled by the cruel Baron Vladimir Harkonnen on Giedi Prime. Herbert's ultimate society is, of course, that of the Fremen of Arrakis. Archaeologists are the cultural anthropologists of the past, studying societies that existed hundreds or thousands of years ago. Herbert pursued this path in anthropology with the bits and pieces of information he provided on how past societies were dominated by thinking machines and how those past societies impact the present. Linguists are anthropologists who study the thousands of human languages, both living and dead, and how language shapes a view of the world. Examples of this in *Dune* include instances of Bene Gesserit code phrases and the speech patterns of various schools of espionage. Physical anthropologists focus on human biology whether it is human evolution or contemporary genetics. Herbert looked at topics ranging from the effects on the human body of geriatric melange to the breeding program of the Bene Gesserit.

Anthropology is best known, however, for its examination of exotic cultures, and anthropologists pride themselves on their ability to interpret other ways of life. In writing *Dune*, Frank Herbert drew most heavily upon anthropology in describing the Fremen people and their traditions. Nowhere is the influence of anthropology on Herbert more powerful than when the Fremen leader Stilgar intrudes upon a strategy meeting Duke Leto Atreides has convened with his staff, early in *Dune* (93–97).

At one point Stilgar pulls aside his veil and pointedly spits on the conference table. Interpreted via their own cultural values, the Duke's men view Stilgar's action as an insult and jump to their feet to restrain the Fremen leader. They stop only because Duncan Idaho, the Duke's Swordmaster of Ginaz, orders them to stop. Playing the role of cultural broker, a translator of other cultures, Idaho assures those at the conference table that they have not been insulted. Instead, he interprets the symbolism of the act in terms of Fremen culture. Idaho reminds them that water is such a precious commodity on the planet Arrakis that, in fact, Stilgar's behavior was an act meant to convey honor, not insult, to the Duke.

Idaho speaks and acts as an anthropologist would: "'We thank you, Stilgar, for the gift of your body's moisture....' And Idaho spat on the table in front of the Duke. Aside to the Duke, he said: 'Remember how precious water is here, Sire. That was a token of respect'" (*Dune* 95).

Even earlier in *Dune* than the scene just described, the influence of anthropology is felt. Anthropologists often describe the feeling of culture shock anyone experiences when encountering a new and different society for the first time, and anthropologists love to tell stories of their own episodes of culture shock. *Dune* begins with a description of a world so different from our own that the reader experiences some cul-ture shock. One struggles to grasp the social organization of this new world and to deal with all the new terms. There are, in fact, so many new terms invented by Herbert for *Dune* that he devoted twenty pages at the end of the book to a "Terminology of the Imperium." The reader, like an anthropologist beginning fieldwork in a new land, flips back and forth to this dictionary to get the meanings of all the foreign words.

I first read *Dune* on the recommendation of two of my anthropology students. And when I recommend *Dune* to someone else, I give the same words of caution those students gave me: You have to read the first part of *Dune* on the faith you will love this book. Otherwise, your struggle to grasp this new world will make you give up on the book. First the culture shock and then the appreciation.

There is a lot that is new in the world of *Dune*. Within the first ten pages, the reader is exposed to a material culture that includes suspensor lamps and gom jabbars, a social organization of royal concubines and fiefs-complete, and a religion with its "Litany against Fear" and the ability to use Voice. Along with the reader, the main character, Paul Atreides, struggles with culture shock: "Paul's mind whirled with the new knowledge" (*Dune* 4).

Later in *Dune* when Paul and his mother, the Lady Jessica, find themselves refugees among the Fremen, they experience culture shock. People most often experience culture shock over common things like new smells, the taste of food, the distance between people in conversation. When Paul and Jessica arrive at Stilgar's home base, his sietch, they are assaulted with strange and unpleasant smells. By contrast, the Fremen are comforted by the "smells of home."

To help Paul in his adjustment, Jessica coughs to get his attention, and says, "'How rich the odors of your sietch, Stilgar....' And Paul realized she was speaking for his benefit, that she wanted him to make a quick acceptance of this assault on his nostrils" (*Dune* 341).

Within hours of their arrival at Stilgar's sietch, they are exposed to stillsuits and reclamation chambers where a body's moisture in the form of sweat, urine, or feces can be recycled. That same day Paul is also confronted with the new responsibility of caring for the widow and children of Jamis, a Fremen Paul has defeated in mortal combat. If Paul and Jessica are to be accepted among the Fremen, they have to overcome their culture shock and adapt quickly to a new and different society (*Dune* 341–347).

In *Dune*, Herbert quickly draws upon several important concepts from anthropology, including culture, the symbolism of language, and holism. When the Bene Gesserit Reverend Mother Gaius Helen Mohiam tests Paul to see if he is human, that is a microcosm of anthropology's concept of culture. According to anthropologist L.A. White, culture is a uniquely human trait best illustrated by humanity's ability to ascribe symbolic, non-inherent meanings to things, events, and ideas. Language is the proof that someone possesses culture in the form of symbolic thought.

An early example of the symbolism of language is Herbert's introduction of the word *kanly*. In telling the Baron Vladimir Harkonnen that "'the art of kanly still has admirers,'" the Duke has chooses a word that conveys a precise meaning. As the Baron says, "'Kanly, is it?... Vendetta, heh? And he uses the nice old word so rich in tradition to be sure I know he means it'" (*Dune* 15).

Anthropology prides itself on being the most holistic of all the disciplines that study people. That means, in being trained as an anthropologist, one studies topics as diverse as ecology, forensics, family life, and archaeology with the assumption that every category of knowledge affects every other category. That holistic point of view is central to *Dune*. Herbert refers to the interconnectedness of everything as a "world's language" and talks of "a world being the sum of many things" (*Dune* 32–33).

The influence of anthropology is obvious in Herbert's creation of the Fremen, a tribal people in danger of being destroyed by a technologically superior society bent on getting at Fremen natural resources. It is a story often told in anthropology. In the case of the fictional Fremen, that natural resource is melange, the geriatric spice, found only on Arrakis. Historically, native peoples have been attacked and displaced to get at natural resources as diverse as gold, silver, rubber, timber, fish, oil, coal, natural gas, uranium, and even water.

The story of the Fremen is not unlike that of the real-life Cherokees, who were forced off their traditional lands in the southern Appalachians by a larger white American society out to get the gold discovered there. At the time of European contact, the Cherokees were the largest tribal population in what is now the United States and initially survived better than smaller groups swept away in the flood of Manifest Destiny. In the early decades of the nineteenth century, the westward expansion flowed first around the Cherokees and then began to erode the fringes of tribal territory. Cherokee lands were agriculturally superior, and white farmers wanted the Cherokees removed. The fate of this Native American group was sealed, however, when gold was discovered on Cherokee land. When removal ultimately became a reality, many Cherokees fled to the sanctuary of remote caves as the Fremen do in *Dune*.

The main reason the Fremen survive is the emergence of a new leader, Paul Atreides in the form of Muad'Dib. The step-by-step process by which Paul comes to lead the Fremen is straight out of the revitalization movement theory of anthropologist A. F. C. Wallace. A society under attack can spiral down to cultural death or reincarnate itself through a revitalization movement. Such movements are typically part religious and part political, as with the Fremen. They usually start with one visionary, like Paul, who through a trance-like state realizes how his group can survive and flourish. The visionary then goes on to become a prophet, communicating his message to larger and larger segments of his society.

If successful, a revitalization movement ultimately transforms itself into a religion and/or political entity, which is what happens in the sequels to *Dune*. Real-life examples include the Longhouse Religion of the Iroquois, begun under the visionary Handsome Lake in the late eighteenth century; the Native American Church, begun under Cheyenne visionaries in the late nineteenth century; and the American Indian Movement, begun under visionaries like Dennis Banks, a Chippewa, in 1968.

The revitalization movement that most influenced Frank Herbert, however, has to be that of the Prophet Mohammed and the origins and spread of Islam. Herbert set his revitalization movement in the same physical environment in which Islam began. He also used terms like *mahdi*, the word for the prophesied redeemer of Islam, for the Fremen visionary, Paul Muad'Dib. At the end of *Dune*, when the Fremen are ready to sweep throughout the known universe, the scene is reminiscent of Islam's spread by the sword throughout the known world.

I think the reason I, as an anthropologist, was drawn to *Dune* and all its sequels was the story of the Fremen. It is an archetypal story of tribal survival in the face of conquest. I often joke that my specialty within anthropology is more specific than politics, social organization, and adaptation, that what I really study is adaptive survival strategies of small groups, or how to win when winning is impossible. The Cherokees, with whom I have lived and worked, were depopulated through disease, warfare, and forced removal. At times they seemed to be on the road to extinction. Like the Fremen, however, they have a knack for survival. In *Dune*, Frank Herbert created a world as complex and intricate as any real culture, and the Fremen are at the heart of his book.

SHARLOTTE NEELY, Ph.D., is an award-winning professor of anthropology at Northern Kentucky University. A native of Savannah, GA, she holds degrees in anthropology from both Georgia State University in Atlanta and the University of North Carolina at Chapel Hill. She is the author of the well-reviewed book *Snowbird Cherokees* and a consultant on the award-winning documentary film of the same title. Dr. Neely has a lifelong love of science fiction and is the author, as Sharlotte Donnelly, of *Kasker*, a novel of anthropological science fiction. She thinks *Dune* is the greatest SF novel of all time.

THE REAL STARS OF DUNE

Kevin R. Grazier, Ph.D.

In Hollywood terms, the stars in the Duniverse are really little more than extras. By examining the real-life counterparts of the stars mentioned in Dune, *however, we can determine which of Frank Herbert's worlds could actually exist, and which ones are mere flights of fancy.*

Jessica returned to the book, studied an illustrated constellation from the Arrakeen sky: *Muad'Dib: The Mouse*, and noted that the tail pointed north. (*Dune* 7)

ISITORS TO THE SAN DIEGO ZOO, arguably the best zoo in the world, are routinely astonished at the variety and sheer number of animals on display. What is less obvious is that the zoo is also a horticultural park and, from a sheer dollar-value aspect, the flora are more valuable than the fauna. A very good case can be made, then, that the plants are the true "stars" of the park. In a similar manner within the Dune Universe, the fictitious planets on which the action occurs are said to orbit real, known stars within the Milky Way Galaxy. With a Duniverse filled with such a variety of fascinating personalities, Machiavellian political intrigues, and strange distant worlds, an examination of the stars, which are mentioned only cursorily in the text, is not an obvious study. Like the San Diego Zoo, however, something in the background can sometimes be equally interesting as that which is obvious.

Astronomers: The Stellar Paparazzi

Astronomers are like paparazzi to the stars of our galaxy. They take pictures of stars, always without their consent, and determine who is hot

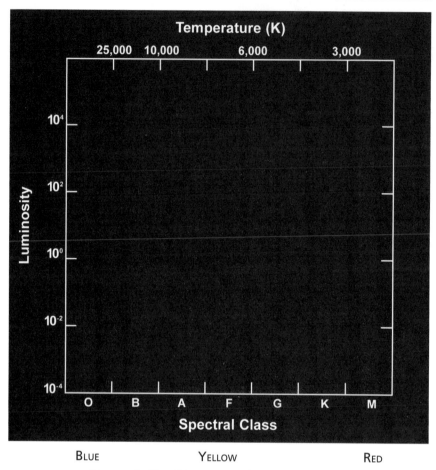

FIGURE 1: THE HERZSPRUNG-RUSSELL DIAGRAM

and who is not. Before we discuss specific stars from *Dune*, let's start with a brief overview on how astronomers classify stars—this will help us to understand why some of these stars are worthy of our interest.

Stars are classified by both their total energy output and their temperature (or color), and one tool used to do this is the Herzsprung-Russell Diagram (or simply the H-R Diagram). Figure 1 shows a blank H-R diagram with no stars plotted, simply to show the framework. The horizontal axis represents the stars' colors or surface temperatures. Surface temperature and color are in some respects the same measure. As an object heats up, it appears to change color. Anybody who has been around a blacksmith or a foundry is familiar with this effect. First, the blacksmith places his cold metal into the fire. As it heats up, it begins to glow red, just like cool stars. The metal gets hotter and it glows orange. As it heats, it then glows yellow and continues through the spectrum of the rainbow until it glows "white hot" (white-hot metal actually has a blueish tinge; very hot stars are similarly blue). In fact, the progression of colors is the exact same as the order of colors on the rainbow: red, orange, yellow, green, blue, and violet. So the color red is associated with cooler stars; blue is the color of very hot stars. It does rather make you question how appropriately labeled are our hot and cold water faucets.

Whereas the X-axis on most scientific plots is displayed with the data increasing in value to the right, on the H-R Diagram the hotter blue stars are to the left and cooler red stars to the right. Stars of different colors fall into different *spectral classes*, ranging from spectral class O for very hot blue stars, to spectral class M for the most common cool red stars. There is, in fact, a mnemonic for remembering the sequence (from hot to cool) O-B-A-F-G-K-M: Oh Be A Fine Girl/Guy, Kiss Me. Alternately, if you wish to remember the sequence from cool to hot— M-K-G-F-A-B-O—then just remember the somewhat more disturbing mnemonic "Mickey Killed Goofy For A Bodily Organ." Whereas our star, Sol, is often called a "yellow" star, the color where it emits most of its light is actually in the green portion of the spectrum. Within each spectral class, there are ten subdivisions, numbered one through ten. To refine further Sol's classification, it is a G2 star.

The Y-axis of the H-R Diagram represents stellar energy output, or luminosity. Light is a form of electromagnetic (EM) energy, as are radio waves, microwaves, X-rays, and Gamma rays. All are different "flavors" of the same type of energy. The total amount of stellar energy output per second is a star's *luminosity*. So a star is classified based upon its color/temperature, and total energy output. The range of luminosities in Figure 1 is based upon the luminosity of Sol. Luminosities in Figure 1 range from 10^{-4} (1/10,000) to 10^{0} (=1) to 10^{4} (10,000) times that of our sun.

A star's *brightness* is a measure of how bright the star appears to the eye (or a telescope), and is a combination of the star's inherent brightness, or luminosity, and its distance from the observer. Within a constellation, stars are ranked in order of brightness and assigned a corresponding Greek letter. For example, the nearest star to Earth is the brightest star in the constellation Centauris and is accordingly called Alpha Centauri. Alpha Centauri is not a very luminous star, but it is bright because it is close. The brightest star in Canis Major, the big dog, is Sirius. Hence Sirius is also called Alpha Canis Majoris. Sirius is both inherently bright and, at only 8.6 light years away, fairly close. The second brightest star in our night sky, Canopus (Alpha Carinae), is quite far away at 310 light years, but appears exceptionally bright due to a very high luminosity.

Stars spend the majority of their lives in a sigmoid-shaped region of the H-R diagram called the *main sequence* (Figure 2), so the vast majority of stars fall into this region. What we see from Figure 2 is that red stars in the lower right are both cooler and dimmer than blue stars in the upper left which are hotter and brighter.

It turns out that there is one more stellar property that is directly reflective of a star's position on the Main Sequence: mass. Small red stars, called *red dwarfs*, have less mass than large blue stars. In fact, the single attribute that determines a star's color, temperature, how long it lives, and how it will die is its mass. This is counterintuitive, but the more mass a star has, the shorter its lifetime. Stars in the upper right of the H-R Diagram live, perhaps, a million years; stars in the lower right may live trillions (the Universe is not old enough for any red dwarf *ever* to have come to the end of its life on the Main Sequence). It does make sense, though, that small stars are far more common than larger stars. M and K red dwarfs like Barnard's Star or Alpha Centauri B comprise 93 percent of known Main Sequence stars. Although our G-type star Sol is often referred to as "average," that term is more representative of the fact that G-type stars fall halfway between the end members of the Main Sequence. Our star is actually larger than 95 percent of known Main Sequence stars. Large blue stars like Rigel are comparatively rare and represent only one in 3 million of known stars.

There are three other notable regions on the H-R Diagram; one represents the stellar graveyard and the other two, stellar hospices—where stars go before they die. We see these regions on Figure 3.

When low- to medium-mass stars near the end of their lifetimes, their outer layers balloon out. The stars become more luminous but actually often cool down. These stars are called *giants*. Eventually

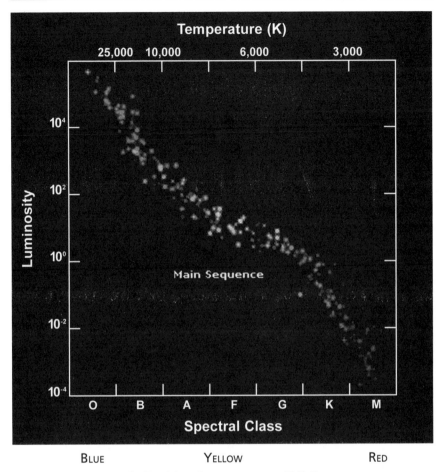

FIGURE 2: THE MAIN SEQUENCE ON THE H-R DIAGRAM

these stars expel their outer shell, creating a cloud of gas called a *planetary nebula*. What is left behind is a very hot stellar core called a *white dwarf*, which takes trillions of years to cool down. When our star enters its red giant phase, the outer layers will expand to beyond the orbit of Earth. The resulting white dwarf will be about the same size as Earth.

Nearing the end of their lives, large stars swell to become *supergiants*, the final region of the H-R diagram. When large stars end their lives, their outer layers are expelled violently in an explosion called a *supernova*. These explosions are so bright that a single supernova can outshine an entire galaxy.

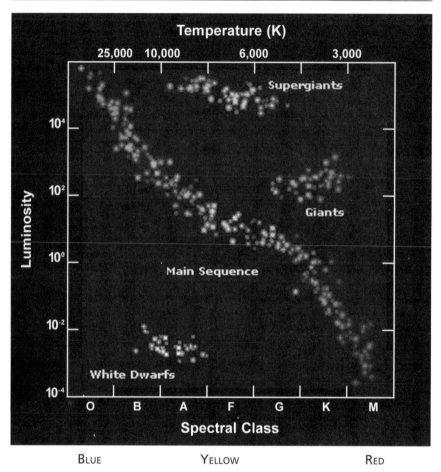

FIGURE 3: GIANTS, SUPERGIANTS, AND WHITE DWARFS ADDED TO H-R DIAGRAM

A supernova seen from Earth in the year 1054 could be seen during the daytime. That explosion left behind an expanding cloud of gas called the Crab Nebula, which can be found between the horns of the bull Taurus. The core of a supernova collapses upon itself to the point where its gravitational field is so intense that protons and electrons are squeezed together to create a form of matter made entirely of neutrons. The name for this material, *neutronium*, finds its origin in science fiction, but has become the accepted term. Neutronium is so dense that a teaspoon of the material would weigh several tons in Earth's gravity. A stellar core made entirely of neutrons is called a *neutron star*. If that core has enough mass, its gravity can be so strong that not even light can escape from it. We refer to that as a *black hole*.

In science fiction, life-sustaining planets are often found orbiting all manners of stars. In reality, though, we can make a good argument that if life, especially intelligent life, is found "out there," it will be around a star very similar to our own. All life on Earth may consume different substances or absorb different gases from the air, but all life requires liquid water. Does life in the Universe necessarily have to be like life on Earth? Does it necessarily require liquid water? Perhaps not, but modern-day science doesn't know about any other forms of life yet, and we certainly do not know how to search for/detect them. Therefore, NASA's search for life "out there" begins with a search for liquid water.

Naturally, the farther one travels from a star, the cooler the temperature becomes. There is a region about every star where we expect that water can exist in its liquid form—it's not too hot, it's not too cold, it's "just right." Understandably, this region is called the *Goldilocks Zone*. For the Solar System, the Goldilocks Zone extends from roughly the orbit of Venus to roughly the orbit of Mars. For larger, hot stars, the Goldilocks Zone is not only farther away, but it is larger than that of Sol's. For smaller, cool stars, this zone is narrower and closer to the parent star.

Now let's examine why we don't think large stars are good candidates for planets with life. Although they have larger Goldilocks Zones, there are other factors that come into play. Earth is 4.6 billion years old and life on Earth, corresponding to the oldest known fossils, began 3.8 billion years ago. That means that it took 800 million years for life to appear on Earth. Stars appreciably larger than Earth's sun live only a billion to a few tens of millions of years. That's not enough time for life to develop. Further, on Earth, complex, multicellular life did not appear until 600 million years ago, so it took 4 billion years to develop. So, while planets orbiting Rigel, a very large blue star, have been a mainstay of *Star Trek* since the original series, in the real Universe, stars like that simply endure for lifetimes that are too short to form habitable planets.

At the other end of the scale we have the smaller, cooler red stars. These stars live extremely long lifetimes, so certainly they're better candidates for life. *Better*, yes, but not *good* candidates. The Goldilocks Zones for small stars are not only very close to the star, they are also very narrow. This presents two problems. First, this dramatically decreases the likelihood that a planet will form in the region that could support liquid water, hence life. More importantly, small stars tend to be more prone to stellar flares. A *stellar flare* is a violent explosion in a star's atmosphere erupting a stream of highly energetic charged subatomic particles into space. If a young planet was in the early stages of creating life-forming compounds, stellar flares could sterilize them, especially since any planet around a red star that could support life would have to be close and would get a bigger dose of radiation from a flare than a planet farther away.

Stars in the middle of the Main Sequence on the H-R Diagram that last long enough for life to form have reasonably sized Goldilocks Zones and aren't as flare-prone as smaller stars. So now that we have a very cursory understanding of stars and, in particular, what kind of stars we might expect to have habitable planets, let's explore the Duniverse.

Dune's A-List

SOL (Earth)

There isn't much to say about either Sol or Earth in the *Dune* series. The surface of Earth was destroyed by atomics during the Butlerian Jihad. Sol, a G-type star, will be around for about another 5 billion years, but life on Earth is extinguished.

CANOPUS (Arrakis/Rakis)

Arrakis—Dune—Desert planet. Canopus—Alpha Carinae—Hot yellow star. In *Dune* we are told that Arrakis is the third planet orbiting Canopus, which is the brightest star seen from Earth's southern sky. In fact, as seen from Earth, not only is Canopus second in brightness only to Sirius, it actually has the highest luminosity—or intrinsic brightness—of any star within roughly 700 light years. Sirius is approximately twenty-two times brighter than Sol, but it is only 8.6 light years away; Canopus is more than 900 times brighter than Sirius.

Canopus is the brightest star in the constellation Carina, so another name for it is Alpha Carinae. Carina has not even been a constellation for very long, historically speaking. The ancient Greek geographer and astrologer Ptolemy identified fourty-eight constellations, fourty-seven of which are still recognized. The largest of Ptolemy's constellations was Argo Navis, representing *Argo*, the ship from Greek mythology used by Jason and the Argonauts, and was simply called Argo. Argo was broken up by the eighteenth-century astronomer Nicolas Louis de Lacaille into Carina (the keel of the ship), Puppis (the stern), Vela (the sails), and Pyxis (the compass). Were Argo still considered a lone constellation, it would be the largest of all eighty-eight.

Canopus is approximately 310 light years from Earth, but due to its unusual nature, its distance was quite uncertain for a long time. Astronomers believe that supergiant stars nearing the end of their relatively short lives can transition from the O/B (blue) region of the H-R diagram to the K/M (red) region and back. This can, perhaps, happen several times. Canopus is a rare, poorly understood, yellowish-white supergiant star—one which may be in a transitional, certainly unstable, state. This state made it difficult for astronomers to estimate its luminosity and so its distance. Until the European Space Agency's Hipparcos satellite made accurate measurements using a property called *stellar parallax*, estimates of the distance to Canopus ranged from ninety-six to 1,200 light years.

Knowing what we know about supergiant stars, we can raise some serious questions about Arrakis. Although Canopus is relatively small for a supergiant star, it is a supergiant nevertheless, with a lifetime of only a few hundred million years. While that's certainly long enough for planets to have formed, it took life on Earth 800 million years. So if Arrakis was a real planet, we could assume that any life there, just like the human life, would not be indigenous! We can go a step further. When the first life on Earth came into being, approximately 3.8 billion years ago, Earth's atmosphere was not breathable to animal life. There was no oxygen. The first life forms, cyanobacteria, inhaled Earth's early atmosphere—gases like carbon dioxide (CO_2), methane (CH_4), ammonia (NH_3), water (H_2O)—and they exhaled oxygen. Initially any exhaled oxygen instantly reacted with the gases present in the atmosphere. Over time, however, the reactants ran out and the level of oxygen in Earth's atmosphere rose. This happened 2 billion years ago and the event is actually called the "Rise of Oxygen." Given what we know about Earth's history and applying that to our search for life in the cosmos, if any planet were discovered to have an atmosphere with appreciable oxygen content, scientists would automatically assume it had life. Arrakis may be hot, but it was certainly depicted as having a breathable atmosphere. Considering the lifetime of Canopus and between the breathable atmosphere and the life forms present on Arrakis, the implication would be clear if it had been a real planet: none of the life on Arrakis was indigenous and Arrakis itself was, in fact, terraformed!

DELTA PAVONIS (Caladan/Dan, Harmonthep)

The original home of the House Atriedes, like many of the important planets in *Dune*, orbits in Earth's southern sky and is impossible to see from most of Earth's northern hemisphere. In the Duniverse, Caladan, then later simply Dan, is the third planet orbiting the star Delta Pavonis, just shy of twenty light years away. The name Delta Pavonis implies that it is the fourth brightest star in the southern constellation Pavo the Peacock. Created in the sixteenth century, Pavo, like many southern constellations, has a more historically recent genesis than many of the northern constellations that date to antiquity, so there is little lore about Pavo.

Like Canopus, Delta Pavonis appears to be in a transitional state in its history. It is a yellow subgiant, appears to be nearing the end of its life on the Main Sequence, and is in the early stages of expanding to a red giant. Therefore Delta Pavonis is cooler and redder than Sol, yet it is slightly brighter.

Terrestrial science has yet to discover Caladan. In other words, no planets have been found to date orbiting Delta Pavonis, but despite being older than Sol and nearing the end of its life, Delta Pavonis is very similar to Sol. Recall the claim that if life is found elsewhere in the galaxy, it will likely be on a planet orbiting a star similar to Sol. This star intrigues scientists who are searching for extraterrestrial life because, in fact, it is so similar to Sol. Scientists at the SETI (Search for Extraterrestrial Intelligence) Institute even ranked Delta Pavonis as the best SETI target, as it is the nearest Sol-like star that is not a member of a multiple star system. Who knows? Perhaps Caladan *does* exist.

36 OPHIUCHI B (Giedi Prime/Gammu)

If we are a product of our environments, then perhaps the complex Machiavellian machinations of House Harkonnen stem, at least in part, from the complex dynamical interactions of their home planet, for Geidi Prime is the fourth orbiting 36 Ophiuchi B. Normally an orange-red K1 star like 36 Ophiuchi B might be a fairly good candidate for an Earth-like planet, but when that star is part of a trinary system, things get a bit more complex.

It turns out that approximately half of the points of light you see in the night sky are multiple star systems, meaning that the lone point you see is actually light originally emitted from two (or more) stars in mutual orbit that has been blended together by distance. Most multiple star systems are binary stars, yet systems are known with three (called trinary systems, like Polaris), four (quaternary systems, e.g., Epsilon Lyrae, also known as the double-double), or with up to six stars (sextuple systems like Castor, the brightest star in Gemini).

Normally when there are more than two stars in a multiple star system, stars tend to pair up. Trinary systems tend to be a binary pair orbited by a lone third star. 36 Ophiuchi, approximately 19.5 light years from Earth, is such a star system. The primary (36 Ophiuchi A) and secondary (36 Ophiuchi B) stars are red-orange K0 and K1 stars respectfully, each having 85 percent of the mass of Sol. They are orbited distantly by a smaller red K5 star.

Recall the memorable scene from *Star Wars* where Luke Skywalker, contemplating his future, watched the twin setting suns of Tatooine (George Lucas's analog of Arrakis). Throughout the years, planets orbiting multiple star systems have been featured in science fiction (e.g., *Nightfall, 2010: Odyssey Two*, and even recent episodes of *Battlestar Galactica*). In reality, though, not all orbits around binary stars are dynamically stable, and planets around binary stars would be confined to a few orbital niches. Computer simulations have shown that planets in a binary system, again remembering that multiple systems are usually groupings of binaries, can exist in two regions. Planets in "p-type" orbit both stars distantly. Planets in "s-type" orbits are very close to one of stars in the binary where the gravitational influence of its central star is very strong, yet the pull of the stellar companion is a weak gravitational perturbation.

If Geidi Prime is the fourth planet orbiting 36 Ophiuchi B, then the implication is that it is in an s-type orbit. Further, if it is the fourth planets, and s-type orbits are close to their central stars, then there are a lot of planets packed into a very close proximity! Certainly in the Duniverse 36 Ophiuchi would display some fascinating dynamical interactions above and beyond the political machinations of House Harkonnen.

ALPHA PISCIUM (Kaitain)

Kaitain is described as the home of House Corrino and, in fact, the capital of the Imperium. The capital was moved to Kaitain after the previous capital, Salusa Secundus, was rendered nearly uninhabitable by atomics:

> Following the nuclear holocaust on Salusa Secundus...everyone had been anxious to establish an optimistic new order. Hassik III had wanted to show that even after the near obliteration of House Corrino, the Imperium and its business would continue at a more exalted level than ever before. (*Dune: House Atreides* 425)

The new capital on Kaitain was described as wondrous to behold:

> On the Imperial planet Kaitain, immense buildings kissed the sky. Magnificent sculptures and opulent tiered fountains lined the crystal-paved boulevards like a dream. A person could stare for hours....Kaitain was exquisitely planned and produced, with tree-lined boulevards, splendid architecture, well-watered gardens, flower barricades...and so much more....Official Imperial reports claimed it was always warm, the climate forever temperate. Storms were unknown. No clouds marred the skies...when the ornate Guild escort craft descended, (Kynes) had noted the flotilla of weather satellites, climate-bending technology that— through brute force—kept Kaitain a peaceful and serene place. (*Dune: House Atreides* 11–13)

In reality, even in the absence of nuclear holocaust, Kaitain is likely to be less inhabitable than Salusa Secundus.

Kaitain purportedly orbits the brightest star in the constellation Pisces: Alpha Piscium. Alpha Piscium is a binary system about 139 light years from Earth. Composing the system is a close pair of a-type stars. In fact, a good analogy would be to imagine the well-known blue stars Vega and Altair in mutual orbit. They would be hot, bright, and each have a lifetime of approximately 1 billion years.

Consider that the life zones around bright stars are larger than those around cooler stars, but are also more distant. Also that planets orbiting in binary systems are either very close to one of the stars or distant and orbiting both. Alpha Piscium is a close binary, meaning that there is little room between the stars for planets and any planets there would be bathed in the light of two very bright stars. That an even remotely habitable planet could exist there is very unlikely. Alternately, a planet in a binary system can orbit both stars distantly. With two very bright stars in close orbit, the Goldilocks Zone may extend to objects in p-type orbits. Then, however, we consider the short lifetime of both of the stars. So, as with Arrakis, any life on planets orbiting this system would likely not be indigenous and habitable. The planet would likely have to be terraformed. Further, at a fundamental level, weather is driven by solar heating, or, more to the point, differences in heating. Were Kaitain orbiting a pair of blueish stars, insulation would be extreme and its weather would be anything but temperate. It is unlikely that technology could change this, brute force or otherwise. Of the planets and their host stars, we have considered so far Kaitain as the least likely as a habitable world.

ALPHA CENTAURI (Ecaz)

Ecaz is not a particularly important planet in the Duniverse. The fourth planet of Alpha Centauri B, Ecaz is useful for its agricultural contributions to the Imperium.

At 4.37 light years distant, Alpha Centauri—also known as Rigil Kentauris—is the nearest star system to Earth. Although only about half the points of light in the night sky are multiple star systems, such systems are well-represented among the important stars in the Duniverse: Alpha Centauri is a trinary system. The system consists of a star slightly larger than Sol (Alpha Centauri A, a G2 star) in mutual orbit with a star slight smaller than Sol (Alpha Centauri B, a K1 star). In orbit about the both of these is the little red M5 star Proxima Centauri.

Proxima Centauri orbits distantly from the binary pair and is only 4.22 light years from Earth—making it the closest single star.

Ecaz orbits a star that, normally, may be quite amenable to life. A K1 star, Alpha Centauri B would be both smaller and cooler than Sol, and its Goldilocks Zone both narrower and closer to the parent star. Alpha Centauri is a trinary system, though. Little red Proxima is so small and so distant that it would not affect the dynamics of the system much. Alpha Centauri A, however, would largely dictate where planets could orbit about Alpha Centauri B. We already know that Ecaz would be in an s-type orbit, because it is stated explicitly in *Dune* that it orbits Alpha Centauri B. By virtue of the fact that the *periastron distance* (the minimum orbital distance) of these two stars is only 11.2 astronomical units (AU) (or 11.2 times the average distance between Earth and Sol), Alpha Centauri A would have swept away any material that might have formed large outer planets like Saturn, Uranus, or Neptune. At closest point of approach, Alpha Centauri A would be 1.2 percent as bright as the Sun seen from Earth. Planets orbiting very close to Alpha Centauri B, therefore, would be largely unaffected by the larger stellar companion. They would neither be dynamically unstable nor would they be fried by the radiation of the larger star.

Because Alpha Centauri B is somewhat smaller than Sol, to receive an amount of sunlight equivalent to that which Earth receives, Ecaz would have to orbit its star at roughly 0.7 times the Sun-Earth distance—or at roughly the Sun-Venus distance. Since Ecaz is the fourth planet orbiting Alpha Centauri B, then there would be a lot of planets packed into a small space. Imagine if Venus were not the second planet orbiting Sol, but rather the fourth. All things considered, Ecaz could, perhaps, be voted "The Dune Planet Most Likely Actually to Exist."

GAMMA PISCIUM (Salusa Secundus)

Salusa Secundus, third planet orbiting Gamma Piscium (Pisces the Fish), has quite a dramatic and tarnished history. It was the capital of the League worlds before and during the Butlerian Jihad. It was, in fact, the home of Serena Butler. Initially, it was described as a near paradise:

Salusa Secundus was a green world of temperate climate, home to hundreds of millions of free humans in the League of Nobles. Abundant water flowed through open aqueducts. Around the cultural and governmental center of Zimia, rolling hills were embroidered with vineyards and olive groves. (*Dune: The Butlerian Jihad* 20)

Devastaed by atomics, the surface of Salusa Secundus was rendered harsh and barren: "'Consider Arrakis,' the Duke said. 'When you get outside the towns and garrison villages, it's every bit as terrible a place as Salusa Secundus'" (*Dune* 29). It was used by the Imperium as a prison planet and as a training ground for the Sardaukar warriors.

Based upon its star, what might Salusa Secundus be like if it were really to exist? Gamma Piscium, 131 light years from Earth, is an orange-yellow giant star. It is leaving the Main Sequence and is likely in the preliminary stages of becoming much larger. It will soon become a red giant in cosmic terms, eventually blowing off its outer layers to leave behind a white dwarf.

Gamma Piscium has, therefore, been expanding for a long time relative to human experience. Any planets that would have been in its Goldilocks Zone while this star was on the Main Sequence would now be hot and likely barren. So it is unlikely that the beautiful Salusa Secundus of the League of Nobles era could exist, but the harsh, barren Salusa Secundus of the Imperium is easy to imagine.

An oddity of this star system, Salusa Secundus would likely present a problem for Guild Navigators. One property that distinguishes Gamma Piscium from stars in its local neighborhood is its relative speed. Stars are expected to move through space with a speed that is, more or less, dictated by their distance from the galactic center. Gamma Piscium is moving more than seven times faster than stars in its local neighborhood. This implies that, like sandworms on Arrakis, this star is not indigenous to this part of space. Analyses of the star's composition confirm that it formed elsewhere and is simply a short-term visitor to this part of the galaxy.

MU BOÖTIS (Ix)

The name Ix comes from the Roman numeral IX, since Ix is the ninth planet of Alkalurops—another name for Mu Boötis. No reference in

Dune is made to the fact that this is a multiple star system, but it is yet another trinary system (and there are hints that it is actually a quaternary system) 121 light years from Earth. Mu Boötis A is a green-yellow F0 star (yes, stars come in green). Based upon peculiarities in the light from this star, however, it may have previously been an undiscovered binary companion. Both Mu Boötis B and C are Sol-like G1 stars.

Since *Dune* makes no mention that this star is a multiple star system, let alone which star Ix may orbit, have we any clues that may help us determine this? All three stars are hotter and more massive than Sol, and although they have shorter lifetimes, all of the stellar lifetimes of the stars in this system are still several billions of years. So stellar lifetimes would have no bearing on the ability of any or even all of these stars to support habitable planets. In fact, since the distance between B and C is 54 AU, and the distance between A and B/C is 4,000 AU, planets could exist around the B/C pair with both p-type and s-type orbits. Planets orbiting within the B/C binary pair would have to remain fairly close to their star, and it is unlikely that nine rocky Earth-like planets could exist in such confines. If there are planets in p-type orbits, they are very distant and cold. Mu Boötis A, however, is fairly hot. Recall that for hotter stars, the Goldilocks Zone is both farther out and larger. So barring a companion star in an orbit that complicates matters, and elaborating on Herbert's *Dune*, Ix is mostly likely the ninth planet orbiting Alkalurops A.

THETA ERIDANI (Tleilax)

The Bene Tleilax, or Tleilaxu, are described in the Duniverse as being quite xenophobic. If they wanted privacy, they picked a planet orbiting the right star for it. Theta Eridani, or Acamar, is a binary pair of blue stars in spectral classes A1 and A4. This star system is, in fact, very similar to Alpha Piscium, the star which Kaitain ostensibly orbits. Clearly, this is not a "life-friendly" star, and would make an excellent "hiding place" for a culture not wishing to interact with others.

40 ERIDANI A (Richese)

Richese is the fourth planet orbiting Eridani A (more correctly, 40 Eridani A). The 40 Eridani system is another trinary, but despite the fact that the star which Richese orbits is stated explicitly, an examination of the B and C members of this system quickly points to the fact that in a search for a habitable planet, there is only one choice in this system. 40 Eridani A is in orange star of spectral type K1 on the Main Sequence and is, in fact, very similar to two other stars we have visited, Alpha Centauri A and 36 Ophiuchi B. Because 40 Eridani A is cooler than Sol, the Goldilocks Zone is at 0.6 AU, closer to its star than Venus is to the Sun.

40 Eridani B and C are in mutual orbit approximately 418 AU from A. 40 Eridani B is a white dwarf, the super-hot core of a star that had ended its life. Presumably, since this star's lifetime has already come to an end, it was the largest and hottest star of the three. We can also presume, that because it left behind a white dwarf, it was not a particularly large star—perhaps a G- or F-spectral class. When B blew off its outer envelope during its red giant phase, any habitable planet around it would have been destroyed or sterilized, yet at 418 AU distance, Eridani A may have suffered few ill effects. 40 Eridani C is an M-class red dwarf, prone to planet-sterilizing stellar flares.

Richese is mentioned in the original Dune trilogy in the glossary of the first book only, hence it would be hard to make the claim that this is an important place. The primary reason that this planet was described here is that in another realm 40 Eridani A is orbited by one of the most famous planets from science fiction: Vulcan.

Chapterhouse

Only the Bene Gesserit know where Chapterhouse is and what star it orbits, and they aren't talking.

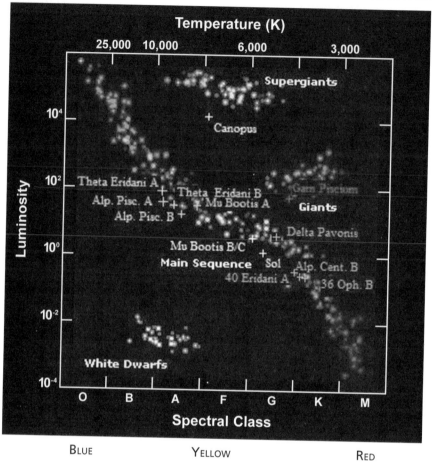

FIGURE 4: H-R DIAGRAM SHOWING PROMINENT DUNE STARS

On the Red (and Yellow and Blue) Carpet

In summary, Figure 4 shows where the important stars, i.e., the stars with really important planets of the Duniverse, fall on an H-R Diagram. Multiple star systems, oddballs, uncommon giants, and rarer supergiants. . . . Clearly the stars of the Duniverse have some very *colorful* characters. Oddly enough, it seems that the most unusual of all these is the *star* of our show: Canopus, home to Arrakis.

Distances are in Light Years	Earth	Arrakis	Bela Tegeuse	Caladan	Chusuk	Corrin	Ecaz	Gamont	Geidi Prime	Grumman	Hagal	Harmonthep	Ix	Kaitain	Pontrin	Richese	Rossak	Salusa Secundus	Sikun	Tleilax
Earth		310	1284	20	387	20	4	72	20	72	178	20	20	139	108	16	174	131	17	120
Arrakis	310		1050	300	501	329	308	378	314	378	363	300	415	312	371	300	278	362	320	225
Bela Tegeuse	1284	1050		1285	1159	1295	1284	1327	1298	1327	1296	1285	1386	1211	1379	1269	1256	1301	1299	1195
Caladan	20	300	1285		394	37	16	88	16	88	189	0	131	141	103	27	175	128	20	111
Chusuk	387	501	1159	394		382	390	383	401	383	508	394	460	258	479	375	526	308	395	346
Corrin	20	329	1295	37	382		23	52	30	52	178	37	108	140	106	30	185	127	22	135
Ecaz	4	308	1284	16	390	23		75	17	75	178	16	122	141	105	18	172	133	16	119
Gamont	72	378	1327	88	383	52	75		77	0	177	88	79	165	116	80	210	144	69	182
Geidi Prime	20	314	1298	16	401	30	17	77		77	182	16	116	152	90	34	173	144	9	127
Grumman	72	378	1327	88	383	52	75	0	77		177	88	79	165	116	80	210	144	69	182
Hagal	178	363	1296	189	508	178	178	177	182	177		189	156	292	189	183	200	305	183	266
Harmonthep	20	300	1285	0	394	37	16	88	16	88	189		131	141	103	27	175	128	20	111
Ix	20	415	1386	131	460	108	122	79	116	79	156	131		241	86	135	200	206	112	241
Kaitain	139	312	1211	141	258	140	141	165	152	165	292	141	241		236	124	287	95	148	101
Pontrin	108	371	1379	103	479	106	105	116	90	116	189	103	86	236		124	186	190	93	207
Richese	16	300	1269	27	375	30	18	80	34	80	183	27	135	124	124		177	128	32	107
Rossak	174	278	1256	175	526	185	172	210	173	210	200	175	200	287	186	177		302	129	138
Salusa Secundus	131	362	1301	128	308	127	133	144	144	144	305	128	206	95	190	128	302		129	138
Sikun	17	320	1299	20	395	22	16	69	9	69	183	20	112	148	93	32	129	129		129
Tleilax	120	225	1195	111	346	135	119	182	127	182	266	111	241	101	207	107	138	138	129	

FIGURE 5: A "MILEAGE" CHART FOR THE DUNIVERSE

Epilogue: Star Maps Sold Here

If the Rand-McNally Atlas survives to the year 13,000 A.D.—or 0 B.G. in Dune terms—then perhaps it would contain the following entry (useful for both Guild Navigators and interested passengers) to help determine, if nothing else, the expected travel time between stars. Imagine the trip from Arrakis to Chusuk: 501 light years' worth of "Are we there yet?"

KEVIN R. GRAZIER, PH.D., is a planetary scientist at NASA's Jet Propulsion Laboratory (JPL) in Pasadena, CA, where he holds the dual titles of Investigation Scientist and Science Planning Engineer for the Cassini/Huygens Mission to Saturn and Titan. There he has won numerous JPL- and NASA-wide awards for technical accomplishment. Dr. Grazier holds undergraduate degrees in computer science and geology from Purdue University, and another in physics from Oakland University. He holds an MS degree in physics from, again, Purdue, and he did his doctoral work at UCLA. His Ph.D. research involved long-term, large-scale computer simulations of Solar System evolution, dynamics, and chaos—research which he continues to this day. Kevin is also currently the Science Advisor for the PBS animated series *The Zula Patrol*, and for the Sci-Fi Channel series, *Eureka*, as well as the Peabody-Award-winning *Battlestar Galactica*.

Commited to astronomical education, Dr. Grazier teaches classes in stellar astronomy, planetary science, cosmology, and the search for extraterrestrial life at UCLA, Cal State LA, and Santa Monica College. He has served on several NASA educational product review panels, and is also a planetarium lecturer at LA's landmark Griffith Observatory.

He lives in Sylmar, CA—and occasionally Mesa, AZ—with a flock of cockatiels and a precocious parrot.

References

Bennett, et. al.,*The Cosmic Perspective 4th Ed.* San Francisco: Addison-Wesley, 2007.

Boss, Alan (2006). "Planetary Systems can form around Binary Stars." Carnegie Institution.
<http://carnegieinstitution.org/news_releases/news_0601_10.html>

Daniels, Joseph M. (1999). *The Stars and Planets of Frank Herbert's Dune: A Gazetteer.*

Herbert, Brian,and Kevin J. Anderson. *Dune: The Butlerian Jihad*. CITY: Tor Books, 2002.

Herbert, Frank. *Dune*. CITY: Chilton Books, 1965.

Herbert, Frank. *Chapterhouse: Dune*. CITY: Putnam Publishing Group, 1985.

Quintana, Elisa V. and Jack J. Lissauer (2006). "Terrestrial Planet Formation in Binary Star Systems" *Icarus*. V 185, 1, 1–20.

Schirber, M. "Planets with Two Suns Likely Common," Space.com, 17 May 2005. <http://www.space.com>

Turnbull, M.C., J.C. Tarter (2003). "Target Selection for SETI. II. Tycho-2 Dwarfs, Old Open Clusters, and the Nearest 100 Stars." *The Astrophysical Journal Supplement Series* 149 (2): 423–436.

Wigert, P.A. and M.J. Holman (1997). "The stability of planets in the Alpha Centauri system." *The Astronomical Journal* 113: 144–1450.

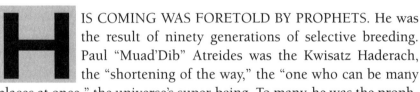

PRESCIENCE AND PROPHECY

Csilla Csori

When you first heard about a book covering the science of Dune, you just knew there would be a chapter on prescience. Csilla Csori knew it long before you did, so we let her examine the chaotic nature and implications of seeing into the future... as well as numerous potential variations.

HIS COMING WAS FORETOLD BY PROPHETS. He was the result of ninety generations of selective breeding. Paul "Muad'Dib" Atreides was the Kwisatz Haderach, the "shortening of the way," the "one who can be many places at once," the universe's super-being. To many, he was the prophet of God. To some, he was the Messiah. His godlike powers included prescience, the ability to see the future. But Muad'Dib did not have vague and cryptic visions of what was to come. He could examine the details of possible futures and direct events toward the outcome of his choosing. How real is this mastery of time, where the future can be seen as clearly as the past? Are there conditions beyond divine inspiration which would allow you to glimpse the future? If yes, do you condemn yourself to a predetermined universe without free will? Or does knowing the future give you the opportunity to change it?

Scientists don't even agree on the definition of time, let alone its nature. It can be described as the relationship between events, i.e. the order in which things happen, but that raises the question of whether time exists when nothing is changing. If time does exist independent of events, then the definition is lacking, whereas if it does not, then some would argue that time itself does not exist except in our perception. And one cannot discuss the nature of time without entering into the realm of religion and philosophy—a delicate area to tread. While scientific and religious beliefs are not necessarily opposed, the underlying assumptions of each stand at opposite poles. How do you know a thing to be true? What is necessary for you to believe that x=y is true? A scientific answer assumes that evidence is key. Not only can evidence prove the truth of the statement, if evidence indicates otherwise, then the statement is assumed to be faulty and a new statement must be sought. A religious answer, on the other hand, assumes the statement to be true based on source, as in doctrine or holy text. If evidence indicates otherwise, then the evidence is assumed to be faulty and new evidence must be sought.

How, then, do you answer questions on the nature of time? Is time like a line, through which your life is a one-way journey, or is it like a circle, wherein the future becomes the past and you are reborn again and again? Your answer will likely fall further into the realm of doctrine than of evidence. Yet you cannot discuss the possibility of knowing the future without first defining what past, present, and future really are and how they relate to one another.

In everyday experience, the future becomes the present, which becomes the past. Muad'Dib described this movement from future to past as the broad vision of the future passing through the narrow door of the present. To him, the future was not a single sequence of events, but rather "countless consequence-lines fanned out" from a single moment. His ability of prescience was twofold: first, to see individual lines of possible futures in stark detail, and second, to know precisely which actions would lead to which future. Whether this holds true in the real world, that the future is an open possibility brought about by your actions, is a question that philosophers have been asking for thousands of years. It is the question of free will. Unfortunately, there is currently no scientific answer to this question, no way to prove that you have a choice of actions and that your future is not predetermined. So you must fall back on the religious answer, to believe in the nature of the future taught by your doctrine. Although some faiths teach that the future is already written, most belief systems allow for some level of free will, even if it is guided by an overarching purpose. You can also assume that you have a choice of actions simply because it feels like you do, or, if for no other reason, because it is the more appealing option. You probably would like to believe that your decisions matter, and that when you choose to make a difference, it is not something that would have happened anyway without you. To believe otherwise is a bleak vision of a pointless existence.

Rolling the Dice: The Probability Model

In real world applications, predicting the future is a matter of probability. If you flip a coin and call heads, you have a 1 in 2 (or 50 percent) chance of being correct. Probability is calculated as the number of correct outcomes divided by the total number of possible outcomes. A slightly less trivial example would be rolling a pair of dice and adding them together, like in craps. There are thirty-six possible outcomes (six possible outcomes on the first die multiplied by six possible outcomes on the second die), and six of those outcomes total 7: (6,1), (2,5), (3,4), (4,3), (5,2), (1,6). So the chance of rolling a 7 is 6/36, or 1 in 6 (17 percent). Calculating probability for dice is fairly simple, because the outcome for each individual die is independent from any other dice, and, unless the dice are loaded, each outcome has the same chance of occurring. Consider instead a roulette wheel, in which the variables interact. The speed of the ball moving in one direction and the speed of the wheel spinning in the opposite direction both determine where the ball will ultimately land. Add in the spin on the ball itself, and the problem becomes even more complex. Furthermore, variables like speed and spin are not neatly divided into discrete units like numbers on a die, so the calculations are mathematically still more complicated.

But both the craps table and the roulette wheel are still rather basic examples when compared to real-life events, because each event is independent from those that came before it. No matter if the shooter just rolled a 7 on the last throw, the chances of rolling 7 on the next throw is still 1 in 6. For an example of dependency, consider a game of blackjack. As each hand is dealt, the cards are removed from the deck, so the next hand has fewer and fewer possible cards from which to draw. As the game progresses, it is possible to keep track of which cards have already been dealt and therefore more accurately predict the next card based on what remains in the deck. (This, by the way, is known as counting cards and will get you thrown out of any casino in Las Vegas.) Understanding how previous events will affect the chances of future events forms the basis for making accurate, real world predictions.

Any system for predicting the future is only as good as the mathematical model it uses. A good model must accurately identify all of the variables affecting the outcome, what possible values the variables can have, how those variables interact with one another, and what all the possible outcomes are, and then determine which combination of variables leads to which outcome. Consider meteorology. It has come a long way since the days of using jars of bear fat to predict the onset of spring, but it can still only tell you that there will be a 40 percent chance of rain in your general area sometime this afternoon, not that it will downpour on your street for twenty-two minutes at exactly 4:31 P.M. Variables include temperature, precipitation, wind speed and direction, humidity, air pressure, cloud cover, warm and cold fronts, ocean currents, etc., all of which interact with one another in complex patterns. Not only are the variables non-discrete—so are the outcomes. Forecasters may say it's sunny, partly cloudy, or overcast, but in truth there are many more possibilities between the extremes of "not a cloud in the sky" and "totally socked in." Sophisticated computer models try to take into account all of the data, but even if weather forecasters were to have a perfect system model, it would only give a definite answer for a limited period of time—a function of the computational precision used to calculate the results.

In our world, only computers can perform these calculations in real time, but in the Dune universe, computers had been outlawed after a crusade against thinking machines and conscious robots. This lack of computers forced human minds to develop instead, giving rise to different training schools. One such school conditioned talented people to perform "supreme accomplishments of logic," which effectively transformed them into human computers, called Mentats. Muad'Dib received the beginning of this training in his youth, and his prescient awareness had its roots in probability mathematics. However, his talent went beyond the abilities of an ordinary Mentat, "as a computation of most probable futures, but with something more, an edge of mystery—as though his mind dipped into some timeless stratum and sampled the winds of the future."

Chaos Rules the World

Beyond the sheer number of variables, the main difficulty in creating an accurate model for predicting weather stems from the fact that weather is a chaotic system, as are most physical systems. Chaotic does not mean random; weather does behave according to specific rules, and there are definite mathematical equations to describe its behavior. A chaotic system, or dynamic nonlinear system, is one in which small changes cause proportionately larger changes, which cause even larger changes, and so on, growing exponentially over time. This is often called the butterfly effect, based on the idea that a butterfly flapping its wings in Brazil can set off a tornado in Texas. This happens because the mathematical equations describing weather are nonlinear.

In a non-chaotic system, described by linear equations, small changes or errors will lead to small deviations. For example, returning to the roulette wheel, if the dealer spins the wheel a tiny bit faster, then the ball will travel a little bit further before dropping. Up until the moment when the ball drops, the system behaves similarly to a non-chaotic system, in that the changes in results are in proportion to changes in the initial conditions. If the ball has the same exact speed and spin in two separate tests, under no circumstance will a small change to the speed of the wheel cause a wildly large change in the distance the ball travels before dropping.

Once the ball drops, however, all bets are off, because the system becomes chaotic. A slight variation in timing of the second drop can cause the ball to strike a ridge and bounce in a completely different direction, with subsequent bounces tracing a path increasingly deviated from the first path. This exponential growth of changes over time is what makes detailed long-range weather prediction impossible with current models. The models cannot take into account every tiny detail of initial conditions. A slight change in one variable will not alter the immediate forecast very much, so it can give an accurate range of a few degrees in temperature. As time increases, however, that small initial change will cause an exponentially increasing range of possible outcomes, until at some point the forecast becomes so widely varied that it is essentially chaotic and unpredictable. It is, in theory, possible that computing power and mathematics will eventually combine to form a perfectly complete system model which measures every detail of every air molecule, but it would still have to account for the final variable—individual decision.

Free will is the most difficult variable to predict in any model. It is not restricted to humans; dogs or even butterflies have some level of decision-making in their actions. According to chaos theory, if you decide that today is the perfect day to go cruising around town, then the exhaust fumes spewed out by your SUV will affect the weather pattern. It won't affect the weather tomorrow or next week, but it might start a monsoon in India six months from now. Any accurate model for predicting the weather far in advance would have to incorporate the effects of individual decisions.

Perhaps the "something more" in Muad'Dib's visions were an ability to predict individual actions with some degree of accuracy. Still, even he was subject to limitations, and he admitted that "a single obscure decision of prophecy, perhaps the choice of one word over another, could change the entire aspect of the future" (*Dune* 218).

The Big Picture: Thresholds, Patterns, and Networks

One emerging field of science which attempts to predict events taking place over large time scales is threshold and pattern dynamics. Rather than looking at everyday events like weather, it examines uncommon and important events such as volcanic eruptions, tsunami, algae blooms, and disease epidemics. These types of events occur suddenly, but they have telltale signs of buildup which often go unnoticed because they take place over long periods of time. Threshold and pattern dynamics attempts to model the factors driving these events using mathematics and computers to identify critical thresholds where the system will shift dramatically. Not only must scientists recognize which factors influence an event, they must develop methods to observe and measure these factors. Then, the next challenge lies in identifying patterns in the observations which indicate where a future threshold will be crossed.

This new field started with efforts to predict earthquakes. For many years, teams of scientists have been collecting data from seismic sensors in earthquake-prone regions. Joining this data with information about geological makeup of the region, they are able to create computer models of stress buildup and movement along fault lines. The enormous number of calculations required for this modeling can only be done on very powerful and fast computers. Today's supercomputers operate in the terabytes (TB) range. (1 TB equals 1,024 gigabytes or over 1 million megabytes. In comparison, the average home computer has 512 megabytes to 1 gigabyte of RAM.) Scientists use this computing power to run detailed computer simulations of earthquakes, such as the TeraShake simulations run by the Southern California Earthquake Center. Using data on how different types of soil and bedrock move in earthquakes of varying depth and magnitude, and how seismic waves travel through these rocks and soils, TeraShake predicted the widespread effects of a future earthquake occurring along the San Andreas Fault. (TeraShake required 1 TB to run, took five days, and output 47 TB of data. It ran on DataStar at the San Diego Supercomputer Center.)

Studies of threshold and pattern dynamics can provide the key information for making these types of projects true earthquake predictors—the key is where and when an earthquake is likely to happen. To do this, a team at the University of California, Davis has developed simulations of entire fault systems, showing the geologic activity over thousands of years. The program simulates the constant rate of tectonic plate movement along a fault, which causes stresses to build up until the rocks slip, triggering an earthquake. By identifying the threshold at which the stress becomes too great, this program can predict when and where an earthquake will occur. Due to the difficulty of modeling complex systems, current simulations can't yet predict exact times and locations of earthquakes. However, they can show that an earthquake is likely to occur in one of a small number of areas and within a specific window of time. The geologic time scale is very long compared to human life, but as the methods of modeling improve and computing power continues to increase, the window of prediction will get smaller.

Research into thresholds holds promise for predicting far more than just natural disasters. It is being used along with network theory to understand the most complicated of systems—those involving people and their individual decisions. Individual actions add up and impact the whole system, making the results impossible to predict in a direct way. Instead, network theory looks at the pattern of connections between people, whether through social or economic ties, and examines how things like disease epidemics (which are the result of the system reaching a threshold) spread throughout the network. For example, studies of the foot-and-mouth disease outbreak in English livestock herds revealed that it was primarily due to an increase in contacts between different herds. Animals were being trucked longer distances, so that farms all across the country became interlinked. The livestock network passed a connectivity threshold, so that once one animal contracted the disease, the infection was able to spread very quickly to other herds throughout the country. This type of study could be used to analyze, and perhaps prevent, the possibility of a global pandemic. While network theory cannot predict an initial outbreak, it can show how quickly a disease would spread, as well as how and if it could be contained. The results of such a pandemic study could indicate where containment efforts would be most effective once an outbreak occurred. Also, if the analysis was to show that the connectivity threshold has already been passed, and that containment would be impossible once the disease started spreading, changes could be made to the network in advance, to prevent any outbreaks from getting out of control.

Whether Muad'Dib had an awareness of threshold dynamics as part of his prescient vision is unclear, but he did recognize when he crossed a threshold of influence within Fremen society and religion. He spoke of a "terrible purpose" that was gathering strength and momentum, and from which he eventually realized he could not escape. He first sensed it as a "wild race consciousness that was moving the human universe toward chaos" (*Dune* 220). By the time he understood what it was—that legions of fanatics would carry out a jihad in his name and that his banner would become a symbol of terror—Muad'Dib also understood that it would continue with or without him. There was no going back from the threshold.

The jihad also acts like an attractor, a state to which a dynamical system will evolve regardless of small changes to the system. Even though the butterfly effect shows that slight changes can lead to vastly different outcomes, certain events are momentous enough that nothing will stop them. These events are constant and unchangeable. An extreme example is the death of our Sun. When the Sun goes nova, it will wipe out everything on the Earth's surface, and nothing that humans do prior to that day will stop this event from occurring. The state of the Earth after that day is an attractor—a constant that we cannot change. Being able to see attractors would give some stability to visions of an ever-changing future. Muad'Dib sensed his "terrible purpose" before he ever set foot on Arrakis, long before he had worshippers or followers. History seems to have been leading toward a jihad, and if it had not been Muad'Dib that precipitated it, someone or something else would have done so eventually.

The Small Picture: Quantum Connections

Muad'Dib described his visions as "prediction of the waveform," which is a reference to a concept in quantum mechanics. Quantum theory attempts to describe the behavior of very small objects (i.e., subatomic particles), which behave differently than larger, everyday objects. This behavior can be observed in a classic physics experiment, the double-slit experiment, which was originally designed to determine if light travels as waves or as particles. In the experiment, light is directed at a barrier with two slits in it, with a detection screen behind it. The pattern the light makes on the screen should indicate if it is a wave or a particle. For example, if the experiment used water instead of light, then a water wave passing through the two slits would cause two sets of ripples to radiate outward and interfere with one another where they overlap. At points where both waves create a crest or a trough, the crest or trough increases, and at points where one wave creates a crest and the other a trough, they cancel each other out and the water remains level. This interference forms a regular pattern on the detection screen. If, on the other hand, bullets were fired at the barrier, then some of the bullets would pass through the slits and strike the detection screen behind it. Those bullets would hit in two clusters, one in line with each of the slits.

When this experiment is performed with light, the results are strange and counterintuitive. Light consists of individual particles, called photons. When photons are fired at the slits one at a time, they build up in an interference pattern of light and dark bands as if they are waves, rather than forming a cluster pattern indicating particles. However, if one slit is closed, then the photons do form a cluster pattern behind the open slit. Each photon passing through one slit behaves as if it is aware of the other slit. Furthermore, if detectors are placed at the slits to record which one a photon goes through (or whether it goes through both at once), then the photons always behave exactly like particles, passing through one slit or the other and forming cluster patterns on the barrier. This holds true whether there is a detector on both or on just one of the slits. Even if a photon passes through a slit with no detector on it, the photon behaves as if it is aware of the detector on the other slit. This duality of particle-wave behavior is not restricted to photons. The experiment has been repeated with electrons and even atoms, which are definitely observable particles with mass, and the results have been the same. Although strange, this behavior is well-documented and is the basis for many real world applications, including lasers, computers, and magnetic resonance imaging (MRI) machines.

The most widely accepted concept which provides an explanation for this odd behavior is the "collapse of the wave function" as described in the Copenhagen interpretation. It posits that what is passing through the slits is a probability wave. Instead of having a definite location, the particle has a probability of being in any particular location. When the photons form the interference pattern of light and dark bands, it is because some locations, the light bands, have a greater probability, and other locations, the dark bands, have less probability. The particle does not really exist as a particle, but as a wavelike property covering the areas where it might be found. The probability wave is spread out and passes through both slits at the same time, with a 50 percent chance of the particle passing through the first slit and a 50 percent chance of it passing through the second slit. These two probability waves recombine with each other on the detection screen and form the interference pattern. That is, until the particle is observed. The particle can only exist as a non-distinct probability when no one is looking at it. Direct observation forces the particle to reveal its actual location, causing the probability wave to collapse into a particle. An observed particle has 100 percent chance of passing through one slit and 0 percent chance of passing through the other. There can be no interference pattern because there is only one probability wave, the 100 percent which has collapsed into an actual particle. So the path of the particle only comes into existence when it is observed. This means that at the quantum level, there can be no outside observer, because the act of observing changes the behavior of that which is being observed.

This concept can be used, as Muad'Dib did, to describe the experience of peering into the future. The future doesn't really exist. It is only a probability wave until the moment when the wave function collapses and the future becomes the present, when probability collapses into actuality. Muad'Dib had the ability to predict the waveform, in other words, to understand and "see" quantum mechanical probability waves without collapsing them. However, his actions were still subject to quantum limitations, and "the expenditure of energy that revealed what he saw, changed what he saw" (*Dune* 296).

The Making of Paul "Muad'Dib" Atreides

Three factors converged in the life of Muad'Dib to give him his special power of prescience: genes, training, and drugs.

By the time of Muad'Dib's birth, the Bene Gesserit sisterhood, an ancient school of mental and physical training specializing in politics, had been running a selective breeding program on the human population for ninety generations. The goal of the program was to breed a new type of human, a super-being they called the Kwisatz Haderach, which translates as "one who can be many places at once." They were trying to create a person with increased mental abilities, who could understand and manipulate the complex mathematics of quantum physics. Such a person would, in essence, be able to think as a perfect system model, calculating all possible variables and their probable outcomes. The Bene Gesserit were breeding for a super-Mentat—a human computer with the prescient abilities found in Guild Navigators.

The Spacing Guild was another of the ancient training schools, one that emphasized pure mathematics. Using a form of prescience, Guild Navigators could "quest ahead through time to find the safest course" for guiding ships through space. Their ability to see the future was limited, however, by nexus points—moments where too many variables compressed too close together in time, so that the results were chaotic and unpredictable. By their nature, navigators always chose a clear, safe path, and they were loath to act when they foresaw a nexus for fear of causing a catastrophe. Muad'Dib could not see beyond a nexus with certainty either, but he was not constrained by the need to choose a safe course, and, in fact, saw that as a path to stagnation.

The combination of Mentat capabilities and navigator-like pre-science would make a powerful and formidable person, one that the Bene Gesserit hoped they could control for their own purposes. They believed that they were only two generations away from their goal, so they instructed Muad'Dib's mother to bear only daughters, one of whom they hoped would bear the Kwisatz Haderach. However, the program was closer to its goal than the Bene Gesserit had calculated, and when Muad'Dib's mother bore a son in defiance of orders, he turned out to be the Kwisatz Haderach they were looking for. Whether they could control him was another matter.

Genetics provided Muad'Dib with a better brain, but that alone would not have been enough for him to realize his full potential and powers. From birth, he was given special Mentat training to develop certain parts of his brain for complex computing. He learned to observe minute details about a person's physiology and behavior, and to analyze those details and draw conclusions about the person's motives, intentions, and future behavior. In addition, Bene Gesserit conditioning gave him precise focus and control over his mind and body. He could replay an event in his mind in slow motion and focus in on particular details. He could control specific muscles and nerve bundles for perfectly timed movement and placement of his body. Using a precise pitch and tone of his voice, coupled with the proper choice of words, he could even manipulate a person into behaving one way or another. This specialized training prepared his mind and paved the way for him to master his powers of prescience.

Even in childhood, Muad'Dib displayed prescient ability in the form of dreams. Not all of his dreams were visions of the future, but he could tell which ones were, and those were correct down to the smallest detail. The catalyst which brought his visions from dreams into the waking world was the mind-altering drug known as melange, or simply, the spice. In small quantities, spice extended life. In larger quantities, it expanded the mind. Guild navigators consumed enormous amounts of spice in order to gain their prescient abilities. As Muad'Dib spent more time on Arrakis and became exposed to greater quantities of spice, he began to have "waking dreams" and visions of different possible futures. For a person with the inherent ability to calculate the future, the spice was creating a positive feedback loop in the pattern-reading portion of his brain.

In a positive feedback loop, the output of a system is fed back in to the input, which accelerates the results. A common example is feedback in a microphone that is placed in front of a speaker. The microphone picks up sounds and the speaker amplifies and broadcasts them. The microphone in turn picks up the amplified sounds and the speaker amplifies them further...until the system overloads and we hear a loud, unpleasant screeching.

A similar loop occurred in Muad'Dib's spice-saturated brain. As the drug expanded his consciousness, he was able to see possible futures. This information was fed back into his mental calculations, which further expanded his ability to predict what was to come. A positive feedback loop has a snowball effect, increasing over time, and, for a while, Muad'Dib's prescience became stronger and more detailed. Obsessed with knowing every aspect of the future, he consumed ever-increasing amounts of spice. However, as with any drug, his body acquired a tolerance for spice, and as a result, he had fewer, dimmer visions. In addition, a nexus would always form around anyone with prescient abilities; the prescient could not see one another with clarity, because the act of looking at the future changed what they saw, and thus the ability to see possible futures made them inherently more difficult to predict. To counteract these problems, Muad'Dib decided to take the Water of Life, a concentrated liquid form of spice, which could cause a permanent consciousness-expanding experience in the user, but which was poison to most people who tasted it. Some Bene Gesserit women were able to consume it and transmute the poison, but all men who had attempted it had died painfully. The Water of Life accelerated the feedback loop in a person's brain to a such a great rate that the person consuming it would either have to alter their brain chemistry to survive, or their brain would overload and break down, like a blown speaker.

Successfully consuming the Water of Life was the final step for Muad'Dib to reach full prescience. The drug amplified the signals in his genetically advanced brain, while his strict and rigorous training allowed him to take control of the process. The tremendous final output resulted in a true mind-altering experience, one which changed his brain chemistry permanently and gave him the power to see past, present, and future with almost absolute clarity. He described himself as "a net in the sea of time, free to sweep future and past. I am a moving membrane from whom no possibility can escape" (*Dune* 506).

Of course, everything has its limits. Even though Muad'Dib could see immediate futures in perfect detail, he could not extend his vision through infinite time. But he could see far enough that, even when he lost his eyes, he could still walk around like a sighted person, playing out the memory of his prescient vision. This feat amazed everyone around him, even those who already believed him to be the Messiah. They marveled that he no longer needed eyes to see, not realizing that he had been living in his prescient vision for a long time. For that is the ultimate pitfall of the oracle: being locked into his vision. Knowing the future with absolute certainty robs a person of free will. Muad'Dib said of himself, "I meddled in all the possible futures I could create until, finally, they created me" (*Dune Messiah* 319). His son Leto, understanding how this could lead the entire human race into stagnation, dedicated his life to undoing it. He took control of the Bene Gesserit breeding program to create humans whose actions prescient people could not predict, thus preserving an unknown and unknowable future.

What Makes a Prophet?

At what point do predictions of the future change from prescience to prophecy? Muad'Dib was more than just a talented man who could calculate quantum physics probabilities in his head. He was more than just a mystic experiencing drug-induced visions of the future. He was a prophet, worshipped by fanatical followers who took his holy war across all the planets of the known universe. Whether or not the hand of God was behind it, millions of people believed that it was. Does a scientific explanation of his abilities necessarily make them less divine?

Many events in Muad'Dib's life mirror those of our own prophets and messiahs. His coming was predicted by other prophets. He was marked as special from the moment he was born. He faced obstacles and trials, including betrayal by a close friend. He led his people to freedom. He passed proverbs and wise teachings on to his followers. And, most notably, he had the power to see the future. How do you determine who is a man with prescience, and who is a prophet of God? Ultimately, his followers decide.

The Guild Navigators had prescient abilities, but they were not revered as prophets. They were part of a business corporation, living separate from any culture or society. Motivated by profit and the preservation of their monopoly, their limited interaction with other people worked only to serve those ends. Muad'Dib, on the other hand, immersed himself in the affairs of the Fremen people. Their fight for freedom became his fight. Before he became their leader, he lived as one of them, embracing their culture and living by their rules. The Fremen considered Muad'Dib to be one of their own, even though he was born an outsider. It is this connection with his people that led the Fremen to follow him with such devotion.

How then, would you recognize a modern prophet? If you met a man with Muad'Dib's extraordinary abilities, would you believe his words to be prophecy? His predictions would be accurate. Is that enough to consider his words to be divinely inspired? Divine or not, would his ability to see the future make you listen to what he has to say? If science could provide an explanation of his abilities, would you be more or less inclined to listen?

Although, by definition, prophecy is divinely inspired, the modern world's various religious beliefs define the concept of divinity in many different ways. If you consider prophecy in simple terms as a statement about the future in which you choose to believe, then what makes a prophet? Does a person have to share your culture or religion in order for you to believe that they are worth listening to? Or is there a universal truth which transcends human differences?

The Dune series shows what can happen when belief in a prophet is taken to the extreme. Muad'Dib tried to escape his "terrible purpose" and prevent the religious war, but ultimately his followers grew into the fanatics he had foreseen, and he could not stop them. In the end, he railed against the religion which gave him godhead and then killed people in his name. But he knew that, even after his death, the jihad would follow his ghost. Muad'Dib was, after all, just a man. In his own words, "There exists no separation between gods and men, one blends softly casual into the other" (*Dune Messiah* 11).

CSILLA CSORI is a programmer/analyst at the San Diego Supercomputer Center. She works primarily on database and software development for business applications, and she also moonlights as a gremlin hunter for her colleagues when their computer programs start acting funny. Recently, she released version 5.1 of ProBook grant application software she authored for the University of California. It's one of those pesky projects that started small but took on a life of its own, and now, like *Doctor Who*'s Cybermen, keeps coming back to demand more upgrades. She gained an interest in quantum physics in college while interning at the Stanford Linear Accelerator Center. In her spare time, she enjoys playing softball, kayaking, and any other excuse to be outdoors in San Diego's perfect weather.

STILLSUIT

John C. Smith

In the harsh environment that is Arrakis, the Fremen owe their existence to their stillsuits. How feasible are stillsuits as they are described in Dune? *John C. Smith examines the ultimate in "green" apparel, as well as the "ick factor."*

A STILLSUIT IS A FULL-BODY SUIT designed to retain and recycle the wearer's moisture. It's powered solely by the wearer's natural body movements. The stillsuit is a convenient *Dune* plot device, since it freed the Fremen society from dependence on water supplies, giving them the mobility needed to thrive in the arid deserts of Arrakis and become a potent fighting force.

The most detailed description of a stillsuit's inner workings was given by the ecologist Liet-Kynes in *Dune*:

"It's basically a micro-sandwich—a high-efficiency filter and heat-exchange system. The skin-contact layer's porous. Perspiration passes through it, having cooled the body . . . near-normal evaporation process. The next two layers . . . include heat exchange filaments and salt precipitators. Salt's reclaimed.

"Motions of the body, especially breathing," he said, "and some osmotic action provide the pumping force. Reclaimed water circulates to catchpockets from which you draw it through this tube in the clip at your neck.

"Urine and feces are processed in the thigh pads. In the open desert, you wear this filter across your face, this tube in the nostrils with these plugs to insure a tight fit. Breathe in through the mouth filter, out through the nose tube. With a Fremen suit in good working order, you won't lose more than a thimbleful of moisture a day—even if you're caught in the Great Erg." (*Dune* 71)

Herbert deliberately skipped over the stillsuit's precise workings. In the engineering world in which I labor, we call this "hand waving," which may be fine for science fiction but leaves a lot of sticky problems left as an exercise for the reader. Some passages seem to indicate that the suit wicks away liquid perspiration from the skin for subsequent passage through filtering layers. In *Children of Dune*, Leto, while in the early stages of his transformation into a worm, noticed that "exertion had produced a slick film of perspiration which a stillsuit would have absorbed and routed into the transfer tissue which removed the salts" (490). References are also made to robes over the suits, which could be used to either reflect sunlight or retain internal heat.

The suit is aptly named since a modern-day still is typically used to separate different liquids by making use of their differing thermodynamic properties. For example, ethyl alcohol (moonshine) can be separated from water and other liquids by heating a mixture of water, sugar, and yeast just enough to boil off the alcohol, but not the water. The alcohol vapor is then condensed to a liquid form. The name stillsuit seems appropriate since the evaporation and condensation of the wearer's moisture is the garment's function, and the suit also separates out salt and other impurities from the wearer's perspiration.

The stillsuit may be the most discussed piece of Dune technology. I know when I first read the book (long before engineering school), I thought they were "kick-ass cool" and made a lot of sense. To be honest, I wanted one and, as demonstrated by many threads on the Internet, many others still do. I was also amazed at how many proposed patents for cooling garments read just like Liet's description of a stillsuit. Herbert or his heirs need not worry, however, since the thermodynamics of most of these patents make little sense.

As fresh water becomes a scarce commodity for much of the Earth's population, and we design spacesuits and habitats needed for future long-term human exploration of the Moon and Mars, the utility of such a "closed" system is truly enticing. But would it work?

Muad'Dib, Paul Muad'Dib

The release of perspiration—sweating—and its subsequent evaporation from the skin is the body's primary cooling mechanism. Failure to sweat leads to heatstroke and death. Recall in the James Bond movie *Goldfinger*, after betraying the evil Auric Goldfinger, Jill Masterson is killed in style by covering her entire body in gold paint. In the movie, the cause of death was incorrectly attributed to asphyxiation, when, in fact, poor Jill would have died from heatstroke. Even when inactive, your body must dissipate about 90 watts of heat just to counter the heat from your metabolism. The skin is responsible for about 90 percent of the body's heat-dissipating function. Enclosing the entire body in a sealed stillsuit, and thereby possibly interfering with the body's ability to sweat, is usually the primary criticism of stillsuit design.

Most of your body's heat is produced in the liver, brain, heart, and muscles. Heat is dissipated by varying the rate and depth of blood circulation within the skin, but the primary mechanism is sweating. Webster's Dictionary defines perspiration as "the transparent, colorless, acidic fluid secreted by the sweat glands." Even when passive, your sweat glands excrete about 0.6 kg of perspiration per day. Maximum perspiration rates can exceed 1.5 liters per hour.

There are two types of sweat glands and each produces different types of perspiration. Eccrine sweat glands are distributed all over the body, but the highest concentration can be found on the forehead, soles of the feet, and palms of the hand. These are the glands involved in temperature regulation. That's why stillsuits have hoods to cover the forehead and also cover the palms and feet. Eccrine sweat is composed mostly of water and various salts (in particular sodium chloride—table salt). Apocrine sweat glands are concentrated in the armpits and in the genital area and their function is less clear, but generally considered to be scent and lubrication. Apocrine sweat contains fatty materials. The main cause of sweat odor is the byproducts of bacteria that break down organic compounds present in apocrine sweat. And yes, as Hawat commented to Paul, the suits "stink to heaven in any closed space" (*Dune* 22).

Would You Drink It?

The stillsuit description states "urine and feces are processed in the thigh pads." Urine is about 95 percent water and the remainder is composed of metabolic wastes (urea), dissolved salts, and organic materials. Incredibly, drinking one's own untreated urine is purported by some to have therapeutic health benefits and the practice dates back to Roman times. However, while drinking small amounts of urine won't kill you (provided it's not infected with germs), drinking your own untreated urine is like recycling your body's waste products. It's just as if your kidneys weren't working and is very hazardous.

While there are many large-scale systems that are capable of recycling urine and other waste water for uses other than drinking, they obviously aren't applicable to stillsuit design. The conversion of urine and exhaled breath back into safe drinking water is being spearheaded by NASA to lessen future dependence on expensive terrestrial fresh water resupply efforts. For example, each kilogram of water transported to the International Space Station (ISS) costs about $5,000 using the Russian Soyuz vehicle and several times more if using the Space Shuttle. The original Salyut and Mir space stations condensed liquid water from the air and used this liquid to produce oxygen for breathing, but the liquid was not clean enough for consumption. NASA's Marshall Space Flight Center is pioneering a method of filtration to produce drinkable water that is consistent with parts of Kynes's description except the device would be huge (bigger than a refrigerator). First, large particles are filtered out by simple filters, and then a carbon filter strips out organic waste products. The liquid is then flushed through a cartridge containing iodinated resins and after prolonged contact with the resins is safe to drink.

Another system under consideration by NASA for the ISS is a vapor compression distillation process in which waste water is boiled, the near-pure water vapor condensed into a liquid, and then purified further. Such devices will be massive but will produce water with contaminant levels lower than most municipal groundwater systems. In the movie *Red Planet*, astronauts portrayed by Val Kilmer and others have some fun peeing with exaggerated urine streams in the low-G Martian environment. While good for humor value, this was a dreadful waste of water.

If you're wondering if processing that other stuff humans excrete (feces) would really be required, the answer is yes, since about 75 percent of human feces is water. The remainder is dead bacteria and indigestible food, such as cellulose. Unlike urine, even small fecal contamination can make you sick—E. coli, present in your lower gastrointestinal tract, does nasty things when it gets into your upper gastrointestinal tract. Not surprisingly, little research has been done on reclaiming drinkable water from feces. However, a study by the University of Colorado on water recycling for a manned Mars habitat concluded that the water reclaimed from feces would be minimal and cited "potential for crew psychological concern." However, for a stillsuit to completely recycle one's moisture, water from feces would have to be reclaimed. The mechanism will gratefully have to be left for future development. The psychological "ick" factor remains but when faced with the alternative of death by heatstroke, it would not be insurmountable.

Inconvenient Thermodynamics

Stillsuits appear to violate thermodynamic principles and would probably not work, at least not in the form described by Herbert. Descriptions of the cooling mechanism in the suit's innermost layer are, at best, inconsistent. It is not clear (and to my mind unlikely) that the body's natural cooling mechanism, evaporative cooling, even occurs. Perspiration must go through the phase change from liquid to vapor in order to remove heat from the skin, thereby lowering the skin's temperature and cooling it. Kynes mentions a "near-normal evaporation process" which is consistent with evaporative cooling, but other passages seem to contradict this.

The method for condensing liquid water from evaporated sweat is not addressed, which is not surprising since condensation requires a mechanism for cooling at least a portion of the suit. Maintaining proper humidity levels near the skin to permit evaporative cooling to continue also requires a cooling mechanism. It's also questionable whether normal movements of the human body would be sufficient to generate the energy required. The lack of a cooling mechanism and sufficient power to supply it are major shortcomings in the stillsuit descriptions. However, possible engineering solutions may exist, and 13,000 years leaves a lot of time for future technology developments.

It's Got to Be a Gas, Gas, Gas

A sealed stillsuit with a porous layer near the skin could reduce or prevent one's sweat from evaporating since it is the evaporation of liquid from the skin that cools it. Any SCUBA diver who's worn a full-body neoprene wetsuit in the hot sun knows how quickly heatstroke can occur. Without evaporative cooling of the skin, or cooling the outer surface of the wetsuit (such as dumping buckets of cold sea water over it), the SCUBA diver would quickly overheat and die if untreated. Ironically, heatstroke is exactly what happened to many of the extras during the filming of David Lynch's movie version of *Dune*. The movie was filmed in the hot Mexican desert and the stillsuit costumes were made out of thin foam which didn't "breathe." Actor Patrick Stewart is said to have commented that the stillsuit was the most uncomfortable costume he has ever worn (and this is the man who invented "the Picard maneuver").

Evaporation is a surface phenomenon in which molecules near the surface have sufficient kinetic energy to change from the liquid to the gaseous state since the molecules aren't constrained in the "up" direction as much by friction. It is a different phase-change mechanism than boiling. For example, a bowl of water will evaporate into the air without the water ever reaching its boiling point. Evaporation always occurs but happens fastest at higher temperatures (higher kinetic energy) and lower humidity, which is why you sweat more on humid days.

Evaporating a liquid such as perspiration requires heat; in other words, evaporation is an endothermic process. This heat is supplied by your body. The removal of heat from the skin cools the skin and the body inside. Dogs, for example, have few sweat glands and thus cool themselves by panting, which evaporates water from the moist lining of their mouths and throats. That's why a wet nose is always a healthy sign for a dog—it's a sign their temperature regulation mechanism is working. Humans who have exceeded their body's ability to cool by evaporative cooling, usually after great exertion such as running, will also pant to provide additional cooling.

In *Dune*, Kynes's description that perspiration (which is by definition a liquid) passes through the inner porous layer after having cooled the skin is consistent with Leto's comments that a stillsuit would have absorbed and routed perspiration from his skin into the outer layers for subsequent filtering. However, just transporting perspiration away from the skin won't cool the skin, it will merely make it dryer. No evaporative cooling occurs, since no endothermic phase change from liquid to gaseous state occurs. Contrary to modern advertisements for high-tech sports apparel—such as workout T-shirts that wick sweat away from your body—these apparel do not keep you cool, merely dry. Wicking garments can be pitched as more comfortable since many people prefer not to be damp, but they are not cooler. Wicking fabrics are great at keeping you warm during cold-weather activities such as skiing because they reduce the amount of evaporative cooling that can occur on your skin.

Wicking fabrics use passive capillary action to transport sweat from the skin surface to the outer surface of the garment and/or to the threads within it. It is on the wicking fabric's surface, not the skin, that evaporation and cooling occurs. Such apparel can theoretically counteract the cooling effects of sweating. In real life this isn't a worry, since sports apparel aren't perfectly skin tight nor are they 100 percent effective at transporting all moisture from the skin's surface. Perhaps the stillsuit's contact layer could be assumed to be highly thermally conductive and engineered such that evaporative cooling does occur. However, such a material does not yet exist (despite some proposed patent claims to the contrary).

If one assumes that a future material can be developed that does allow evaporative cooling of the skin, another problem remains due to the high humidity in that contact layer. The humidity in the porous layer next to the skin would be extremely high in a sealed suit and, if not reduced, would result in even more perspiration in a futile attempt to cool down the body. Somehow the relative humidity of the air in this layer would have to be continually reduced or the vapor surrounding the skin would quickly become saturated, and evaporation would slow to a crawl. The stillsuit would somehow have to transport high-humidity air away from the skin and substitute dryer air at a comfortable temperature in its place. Basically, the air needs to be dehumidified for evaporative cooling to continue.

A boundary layer would also have to exist in the stillsuit between the inner low-humidity layer and outer layers of higher humidity. Such a layer would not be passive. The natural tendency would be for uniform humidity and temperature within the closed suit system in the absence of work, so we must assume that work (energy) is provided to keep the air in the innermost layer of the suit dry (and at a comfortable temperature) while also transporting vapor to the outer cooler layers. Herbert makes no mention of energy used for this task.

Luke Skywalker's Day Job

Condensation is the opposite of evaporation—it's the phase change from the gaseous state to the liquid state. Lack of a mechanism to cause condensation is a shortcoming in all stillsuit descriptions. Heat is released onto the surface where condensation takes place—an exothermic process. For example, if you observe a glass of your favorite beverage with ice on a warm day, water condenses out of the air onto the surface of the glass, warming the glass in the process. However, condensation will only occur if the glass is cooler than the ambient air.

The windtraps of Arrakis collect water by the same method. This points to a challenging problem for stillsuit design. In order for condensation of evaporated sweat into liquid, drinkable form to occur, some fraction of the outer layers of the suit must be cooled.

In Kynes's description, he states, "The next two layers...include heat exchange filaments," but Herbert never details a mechanism for such cooling or really even hints that such refrigeration is necessary (*Dune* 8). Water vapor passing through the stillsuit's nose plugs as the wearer exhales would also have to be condensed for recycling. It would make sense to combine the water vapor passing through the nose-plug tubing with the water vapor evaporated from the skin. This could be accomplished by appropriate routing of the suit's plumbing. However, carbon dioxide from the breath would first have to be filtered out, perhaps by forcing the exhaled vapor through a chemical-based carbon dioxide scrubber canister like those used in spacesuits.

Dehumidification of the stillsuit's innermost layer is needed for evaporative cooling to continue. This process also relies on condensation. A typical room dehumidifier works by forcing the room-temperature air over very cold coils using a fan. Moisture from the air condenses onto the surface of these cold coils and the liquid water drips into a pan or down a drain. Removing water from the air lowers its humidity level. In your refrigerator, as air is cooled after passing over cold copper coils, it is also dehumidified, since liquid water condenses on these coils. These coils are part of the refrigerator's "evaporative cooler."

In *Star Wars: A New Hope*, condensation and dehumidification were also the bane of young Luke Skywalker's existence. On Tatooine, before becoming a Jedi, Luke's job on his Uncle Lars's moisture farm was fixing droids and evaporative coolers. The function of the evaporative coolers was to extract liquid water from the arid desert air and must have operated similarly to a dehumidifier or refrigerator. On desert worlds, even in a galaxy far, far away, water reclamation is an important part of life.

I'd Like a Heat Pump in X-Large Tall Please

Herbert makes no mention of a mechanism for refrigeration of any part of the stillsuit, though he does understand that heat exchange of some kind is required. Firefighters and military personnel who must wear heavy protective vests in hot situations sometimes place "blue ice" packs near the skin to keep the body from overheating. However, assuming that a person traveling in the deserts of Arrakis has access to a resource of something like ice as often as needed seems impractical and inconsistent with the general concept of the stillsuit. For example, Liet's description that the suit would function "even if you're caught in the Great Erg" doesn't have the caveat that you need anything other than the stillsuit (*Dune* 8). Nighttime temperatures in the desert would be cool, but I know of no practical means of extracting and then storing sufficient cooling reserves for daytime retrieval.

There are many methods of refrigeration available. A heat pump is simply a mechanism that moves heat in a direction that provides cooling on a hot day and therefore requires that work (energy) be added to the system. Common examples would be a refrigerator or air conditioner that employ a vapor-compression cycle of a refrigerant and are driven by electricity. Another method is absorptive refrigeration, where the cooling is produced in a somewhat similar manner, using two fluids and heat input instead of electricity. Examples include a camper refrigerator powered by burning natural gas, or the ice factory constructed in the middle of the jungle by Harrison Ford's eccentric character in the movie *The Mosquito Coast*.

Determining the optimal refrigeration method for a stillsuit is a difficult and pointless task since inevitable improvements in technology would shift the outcome. Designing a functional stillsuit would require the efforts of teams of engineers for many years (designing a spacesuit is simple by comparison). Even if a feasible stillsuit could be designed, the technology to make them actually fit comfortably is probably a goal left for the distant future. However, the basic thermodynamic principles and concepts won't change, so let's assume the needed cooling is produced by something you are already somewhat familiar with—your refrigerator.

Thermo 101

First, some basic principles. Heat freely flows from hot to cold. To cause heat to flow from cold to hot, work must be added to the system. An example of work would be your refrigerator's compressor that is powered by electricity. An imperfect analogy is that piped water freely flows downhill but must be pumped (work supplied) to flow uphill. The second key concept is that decreasing the pressure of a gas decreases its temperature. This makes sense, since temperature is a measure of the kinetic energy of the atoms and molecules of the gas as they move and collide. For the same amount of gas, if the volume is decreased, the frequency of collisions increases and thus temperature of the gas increases. Conversely, an increase in volume decreases pressure and temperature.

Pressure and temperature determine the boiling point of a liquid. For example, in a butane lighter, the butane is stored in liquid form under high pressure at room temperature. If you hold the release open, butane released into the air immediately boils into a gas—which can be ignited—even on the coldest of winter days, since butane has a very low boiling point at sea level atmospheric pressure. The lighter will even get cold to the touch since heat is absorbed from the immediate surroundings as the butane boils. A fire extinguisher gets cold for the same reason.

How a Refrigerator (and Stillsuit?) Cools

A refrigerator cools by the removal of heat resulting from boiling a refrigerant (such as Freon) at a low pressure, and the release of heat resulting from the condensation of the refrigerant at a higher pressure. The basic four-step cooling cycle is depicted in Figure 1.

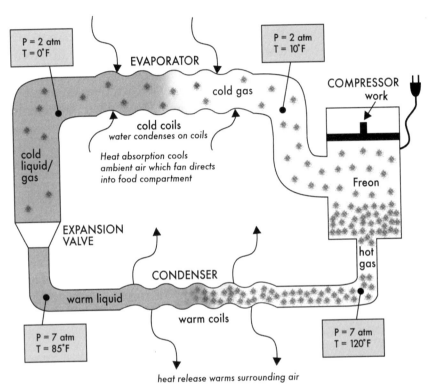

FIGURE 1

The compressor (the mechanism responsible for the rumbling noise your refrigerator produces) circulates Freon in both liquid and gaseous states through wiggles of copper tubing which wind both inside and outside a refrigerator. The copper tubing system depicted in the figure is closed and the plumbing contains a refrigerant. The most common refrigerant used to be the ozone-destroying chlorofluorocarbon Freon but now tends to be a more ozone-friendly hydrofluorocarbon such as R-134a. Let's assume Freon for this example.

Let's examine the cycle starting with the compressor. The input to the compressor is cold, low-pressure Freon gas. Note the pressure is still greater than the ambient air in the room (assumed to be the pressure at sea level = 1 atmosphere = 1 atm). The compressor acts like a piston in a car engine compressing the Freon gas, raising its pressure and therefore its temperature. Remember the last time you pumped up a bicycle tire, the tire got warmer. The output from the compressor is a high-pressure, super-heated Freon gas. The work in this system is done by the compressor, which requires electrical power to function.

The Freon gas then passes through a condenser, which acts as a heat exchanger much like your car radiator, and within its loops, heat escapes from the gaseous Freon to the ambient air surrounding your fridge. The Freon condenses within the tubing into a lower temperature liquid. Pressure is unchanged. In order for heat to flow freely from hot to cold (without work), the system must be designed such that the condensation temperature of the Freon gas is higher than room temperature. That's why the tubing (the condenser) on the back of your refrigerator feels warm to the touch. Note the condenser portion of a stillsuit would cause the outside of a suit to glow in the infrared creating an easily visible target for any enemy with an infrared scope.

The liquid then passes through an expansion valve, which causes the Freon pressure to drop dramatically. Think of a whistling teakettle. This pressure drop causes partial vaporization of the saturated Freon liquid almost instantly. About half the liquid changes phase to gas. Technically this complex process is called Joule-Kelvin expansion, but I prefer (somewhat incorrectly) to think of the liquid as being just on the verge of boiling and the increase in volume just tips the scale toward vaporization. No worries since there are no quizzes in this book.

The liquid/gas mixture then flows through the evaporator section. Room-temperature air is circulated by a fan (that other annoying noise from your fridge) over the evaporator's extremely cold wiggled tubing. The Freon in the tubing absorbs heat from the blowing air and boils completely into a gas, raising the Freon's temperature. No change in pressure occurs. Note that Freon boils near 0° F at the pressure in the condenser. Just as a butane lighter gets colder due to the boiling of the butane, the outside of the copper tubing gets colder due to the boiling of the Freon within removing heat from its surroundings. Some of the blowing air condenses on the cold evaporator tubing and this water dribbles into a pan (and often onto the floor in my case). The blowing air is now cooler and has lower humidity and the fan continues to circulate this cooled air throughout the refrigerator. The cold, low-pressure Freon gas is drawn into the compressor, which then increases the pressure of the gas and continues the cycle. Heat pumps can also be designed to provide heat which might be useful in providing warmth during colder nighttime desert temperatures.

Can the Human Body Power a Stillsuit?

A stillsuit is powered by "motions of the body, especially breathing" (*Dune* 84). However, the energy provided would almost certainly be insufficient to power the cooling mechanisms needed for condensation, dehumidification, and temperature control. The lack of sufficient energy is perhaps the biggest engineering problem overlooked in the stillsuit descriptions but is also probably the easiest one to fix. Since there is always a driving need for more efficient, smaller, and lighter weight power-generating devices, assuming greatly improved means of generating power 13,000 years in the future is probably a good bet. However, augmentation with non-human sources of power, such as solar power, would likely be required.

Deriving power from normal human movement is under study and shows some promise. An MIT study, "Parasitic Power Harvesting in Shoes" (got to love that title) calculated that up to 67 watts of power could be theoretically produced from heel strikes during a brisk walk. Recall heel pumps are used to circulate moisture through a stillsuit. The study noted that extracting this much power from walking would greatly interfere with one's gait, and the control of one's walking motion is important on Arrakis to avoid attracting sandworms. The actual shoes created for the study minimized gait interference and only generated less than 1 watt of power.

A backpack generator designed by Lawrence Rome of the University of Pennsylvania was able to convert the vigorous strides of hikers into 7 watts of power. The up-and-down motion moves a toothed rod which is meshed with a gear wheel, causing it to turn and generate electricity. The inventor credits his first introduction to generating power from ordinary human movements to the stillsuit design in *Dune*. I even own a flashlight that you can charge just by shaking it, but it's tiring and certainly not a normal human motion. Another common example is a wristwatch wound by arm movements.

An interesting technology under development by IntAct Labs in Cambridge, Massachusetts, is using motion-sensitive proteins to generate power. A protein called Prestin, found in the inner ear, converts electrical voltage into motion. The process can theoretically also be reversed, converting motion into electrical power. Networks of the proteins could form "power skins" to coat stillsuits or the robes often worn over them by the Fremen.

Solar power is another attractive additional energy source for a desert world. About 1,300 watts per square meter of solar radiation reaches the Earth's surface. Current technologies can only convert a fraction of this sunlight into electricity but this is another area of rapid technological development. Perhaps photovoltaic elements could be part of the suits' exterior and robes. Batteries would be required to store the power for nighttime usage. Advanced battery technology powers the Fremen paracompass device used to navigate the surface (using local magnetic anomalies as landmarks) so the Fremen already carry one type of compact battery with them at all times. Another power option based on designs of space probes to the outer planets is to generate power by heat from the decay of radioactive isotopes. For example, the New Horizons spacecraft en route to Pluto generates 240 watts using this method. However, strapping on a heavy backpack laden with Plutonium-238 makes this option less attractive.

Bottom Line

Stillsuits designed using strict literal interpretations from the Dune books probably would not work and most likely would cook the wearer like a crock-pot. However, engineering solutions can be envisioned for all of the suit's shortcomings—it's just of matter of when the technology will be available and if the end product would look anything like a stillsuit.

"Need" is the mother of invention, and right now there isn't a driving need for a suit that completely recycles all of a person's moisture.

As fresh water becomes a scarcer commodity, technology to recycle waste water will continue to advance, but nothing is driving it to be miniaturized to the point where it could be incorporated into a suit. It seems likely that water reclamation will eventually be incorporated into spacesuit design, but not until the more daunting limiting factors of breathable atmosphere and power are greatly improved.

Herbert is deliberately vague on the subject of heat exchange, but a stillsuit would need some mechanism for cooling to condense evaporated sweat and exhaled breath and keep humidity and temperature in the suit under control. Artificially produced power supply systems to augment or replace the human body's contribution also seem needed, but financial incentives are leading to great technological strides in this area. The Fremen were not a technologically challenged people—even Duke Leto praised their suit designs as examples of "good engineering." Given that the Dune stories are set 13,000 years in our future, sufficient time exists to develop all the required technologies necessary for a working stillsuit, but it's unclear whether the suits could be made to look and function like those in *Dune* or whether there will ever really be a need for them.

JOHN C. SMITH has a master's in celestial mechanics (the motion of heavenly bodies) and has spent more than twenty years at NASA's Jet Propulsion Laboratory in Pasadena, CA, designing missions to explore the planets. He has worked on successful missions to Venus, Earth, and Mars and has been part of the Cassini/Huygens mission to Saturn and Titan for sixteen years. He is the designer of the four-year "tour" of the Saturn system which began in 2004 and recently contributed to the design of a two-year extended mission through 2010.

THE BLACK HOLE OF PAIN

Carol Hart, Ph.D.

Heart attack victims often feel pain "referred" to their left arm. Amputees feel "phantom pain" in lost limbs. Pain is obviously a complex association of phenomena. Carol Hart, Ph.D., details the mechanism of pain, as well as its transcendent effect on Paul Atreides.

From the folds of her gown, she lifted a green metal cube about fifteen centimeters on a side. She turned it and Paul saw that one side was open—black and oddly frightening. No light penetrated that open blackness. (*Dune* 7)

NDER THE COMPULSION of the Bene Gesserit Voice and the poisoned gom jabbar needle poised at his neck, Paul is forced to put his hand inside the blackness of the green metal cube and to find out what is inside it— a burning pain slowly mounting to an intensity so terrible that he is certain his hand has been reduced to charred bones: "Reason told him he would withdraw a blackened stump from that box.... He jerked his hand from the box, stared at it astonished. Not a mark. No sign of agony on the flesh" (*Dune* 10).

The most astonishing feature of Mother Mohiam's little box o' pain, as Paul instantly recognizes, is the absence of any mark of trauma or even of any residual throbbing. His left hand is aching and bloody from being clenched in his agony, but not his right. "Pain by nerve induction," is the only explanation offered by Mother Gaius Helen Mohiam, as she slips the box back into the folds of her gown. The incompleteness of the explanation is emphasized by her next words: "There're those who'd give a pretty for the secret of this box, though" (*Dune* 10). Baron Harkonnen and his Mentat assassin Piter would no doubt be among the most interested, but readers of *Dune* would also like to know how the box works.

How We Feel Pain

"What's in the box?"
"Pain." (*Dune* 9)

To understand Mohiam's box, we first need to understand what it contains—Paul's hand in a state of pain. The hand is exquisitely sensitive to painful and non-painful stimuli, as torturers throughout the ages have recognized. Think back to all the times you've hurt your hand. Burned fingers and knife cuts are universal experiences. Then there is the bruising pain of a rapped knuckle, a finger pinched in a door, or even an overly aggressive handshake. The numbness and tingling experienced when your hand goes to sleep can be painful if prolonged. Maybe you once blundered into a patch of poison ivy or got sloppy with a bottle of drain cleaner or bleach. Sticking your hand into ice water or a snow drift to retrieve something is fairly unpleasant as well.

With the exception of strong mechanical pressure (bruising, crushing), most of these injuries will trigger identical pain-signaling receptors (known as *nociceptors*) located on sensory nerve terminals—in other words, nociceptors detect injury without discriminating its type or source. While specific temperature receptors can discriminate different ranges of warm and cool temperatures, extremes of either heat or cold trigger the same nociceptors on the same sensory nerve pathways. Because Paul cannot see his hand in the box to know what is happening to it, the intense burning sensation he experiences could be almost anything, including extreme cold.

The progression of sensations is very precisely described. At first Paul feels only a sense of cold and then slick metal as he puts his hand into the utter blackness of the box. (This initial impression of "slick metal" [*Dune* 7] apparently results from casual contact with the interior walls; there is no suggestion that his hand is forced into contact with anything while it is inside the box.) This is followed by a prickling "as though his hand were asleep" (*Dune* 7), then an increased tingling that progresses to an itch, followed by a faint burning sensation that slowly, steadily, intensifies to an extreme level.

144

Think back again to those pinched, cut, burned fingers of yours. There was the initial sensation—the one that made you yank your hand back (too late!) from further injury. That first one-tenth-second response is due to signals traveling on fast pain fibers (called type III or A-delta fibers). The pain of that initial instant isn't really all that awful—just a quick, sharp pricking or stinging sensation. The worse is yet to come, alas! The slow pain fibers (called type IV or C fibers), which need a full second to get their message to the brain, now kick into action. Their stimulation results in an aching or burning pain that spreads, persists, and (usually) worsens with the body's inflammatory response to injury. The interplay of the fast and slow pain fibers is referred to as the "double pain" phenomenon.

The persistence of pain is due to the peculiar obstinacy of C-fiber pain pathways, which differ from other sensory nerves in their slowness to adapt to repeated stimulation. Think of all the tactile stimulation your sensory nerve receptors are getting right now—the pressure of sitting on a chair, the feel of the book in your hands, the contact of your clothing and perhaps a wristwatch, glasses, or jewelry. There is likely some degree of background noise as well—machinery humming, cars passing, a radio or TV playing. If you are interested in what you are reading (as I hope you are), all these ongoing sensations are below awareness, except any that happen to be uncomfortable or mildly painful—from clothing that is too tight or chafing, or the pressure pain of sitting still for too long. Other ongoing sensory signals fade out, but the pain pathways insistently continue to signal and will shut down only slowly after the irritant is removed or the injury has healed.

Under some experimental settings, repeated or ongoing stimulation amplifies pain signaling so that a mildly noxious stimulation—say, a warm electrode briefly applied to the skin along a single nerve pathway—becomes increasingly painful with repetition, a phenomenon known as *pain wind-up* or *temporal summation*. If our other sensory nerves behaved like this, we would be so bombarded by sensation that we could never hold still; the "white noise" hum of machinery would become deafening, and the gentle pressure of clothing on the body would build until it goaded us to madness.

This building of pain with repeated nerve stimulation sounds very much like Paul's experience with Mohiam's box, with one important difference. Whether it is a bumped elbow or a bad burn or a pain wind-up phenomenon, our ordinary pains never shut off abruptly, yet this is Paul's experience. The pain passed instantly from extreme to nonexistent, "as though a switch had been turned off" (*Dune* 9).

Pain Happens in the Brain

"But the pain—" he said.
"Pain," she sniffed. "A human can override any nerve in the body."
(*Dune* 10)

The classic high school textbook presents sensory nerves as though they were telephone wires setting off buzzers in various response centers in the brain, which pick up the line and send a message back. In fact, the intensity of the nerve stimulus is just one (relatively minor) aspect of the complex experience of pain. There is also the emotional and intellectual processing of the pain, which includes the significance the individual gives to the injury and its predicted consequences. In behavioral studies, people report lower levels of pain emotion for the same degree of physical pain when experimenters reassure them (for example) that their hand will not be burned by the stinging hot water bath in which it is immersed. Consider "Damn, cut myself again!" versus "Oh, oh, I'm bleeding, I'm bleeding! I'm going to be sick!" The injury and the physical pain are the same, but one individual feels only annoyance while the other's suffering is heightened by panic and fear.

The difference is not stoics versus wimps. Suppose you obstinately or impatiently chose to pull a searingly hot dish out of the oven with only a thin dish towel to protect your hand—ouch! But you have only yourself to blame. Next, suppose instead that some vicious bully forced your hand down onto a hot stove while you struggled in vain. That would "hurt" a great deal more, wouldn't it, even if the injury to your hand was exactly the same? Psychologists refer to this issue as *locus of control*: either internal (you are able to cope) or external (you feel helpless). Another powerful pain modulator is *attention*. If, despite your badly burned hand, you are focused on urgent business, such as putting out a kitchen fire, you will feel little or no pain, a response called *emergency analgesia*.

It is clear that every feature of Mohiam's test serves to magnify Paul's pain emotion. It also matters very much to his pain and terror that Paul's hand has disappeared into the utter blackness of the box. If he cannot see it, how can he know what is happening to it—or if it is even still there? We can have pain in the absence of tissue injury because of the pain wind-up phenomenon; we can even have pain in the absence of tissue. *Phantom limb pain* is a common affliction of people who have undergone an amputation. These pains were originally thought to be a result of trauma to the severed nerves of the stump, but attempts to treat by a second nerve-sparing amputation frequently made them worse. The more extreme surgery of severing the sensory nerves at the spinal cord also proved to be ineffective in most cases. Pain doesn't happen in the hand, arm, or leg. It happens in the brain.

Technology or Witchery?

"Slowly, feeling the compulsions and unable to inhibit them, Paul put his hand into the box." (*Dune* 7)

Not too shockingly, the Pentagon can envision uses for Mother Mohiam's device and is "paying a pretty" of your tax dollars to study the concept of causing pain without inflicting injury. According to reports on the Web site technovelgy.com and in *New Scientist*, they are working on at least two trauma-free pain-producing weapon systems. The Orwellian-sounding "Active Denial System" uses microwave beams to inflict thermal pain in order to disperse crowds. Unlike Mohiam's box full of pain, anyone who could not escape the beam would soon suffer burns from repeated hits. Another proposed system of "pulsed energy projectiles" would utilize a femtosecond laser beam burst that, on contact with flesh, would generate an extremely brief plasma pulse. In theory, this laser could be calibrated to cause a burst of intense but transient pain without tissue damage. In theory only: with either weapon, any prolonged or repeated exposure would inevitably cause tissue injury. Neither weapon offers a certain parallel to Mohiam's box, which delivers a slowly mounting (and abruptly ending) pain sensation. But, tens of thousands of years in the future, might someone perfect and miniaturize a laser device that would selectively stimulate nociceptors to trigger extreme pain sensations (perhaps by harnessing the pain wind-up phenomenon) without tissue damage and without inflammation? It might be done, but it seems unlikely that the Bene Gesserit would be the ones to do it.

The Bene Gesserit dislike technology and avoid its use. Machines are a crutch, as Mohiam tells Paul (*Dune* 11–12). The Bene Gesserit school is focused on training the human mind and body to a highly advanced state that allows them to penetrate and manipulate the minds and motives of less advanced humans. We might speculate that the box conceals, for example, an intelligent laser device that can isolate and focus on a specific nerve path, even though the hand is apparently free to move within the box. However, at 15 centimeters (six inches), the metal cube is barely large enough to enclose a small human hand. It seems unlikely it could conceal complex circuitry, particularly since there is no suggestion of thick or double walls to the tiny box. But what then could have produced the intense burning sensation that Paul experiences? In a book that is in part about the politics and ecology of energy (where CHOAM can be equated to OPEC), Herbert surely would not invent a fantasy object that required no power source whatsoever.

There is Bene Gesserit witchery at work here. The small box does have an energy source: Paul's hand, which is pumping out infrared (thermal) radiation. Obviously, we do not ordinarily burn ourselves with our own body heat, but we might if it were efficiently collected and reflected within a small space.

On first sight, Mohiam's box would seem quite ordinary—a small, green metallic cube—except for one feature so striking that Paul is immediately frightened. That is the utter and unnatural blackness of its interior: "No light penetrated that open blackness" (*Dune* 7). Paul again describes the cube's interior as "a lightless void" (*Dune* 10) as he obeys Mohiam's order to remove his hand. This emphasis on the utter blackness of the interior suggests it might be a perfect or near-perfect black body—that is, an object or surface that perfectly absorbs all electromagnetic radiation reaching it.

A perfect blackbody absorber will also be a perfect emitter of infrared radiation. The heat of Paul's hand (he is nervous and sweating) is absorbed and emitted to heat the interior (not the green exterior) of the cube. The green exterior would reflect infrared (think of the coolness of grass), so the empty interior would stay cool, even though Mohiam has been carrying the box under her robe, close to her body.

Skin temperature is normally around 90° F; the pain threshold is not drastically higher, only about 113° F. Paul, of course, feels extreme burning pain but suffers no injury at all. Here we need to remember that the complex experience of pain happens in the brain, not in the body's tissues. Paul cannot see his hand but he has been told that the box contains pain. He is surprised when the initial cold of the box begins to give way to warm sensations and then to mild burning. He is fighting down his terror of the box and the gom jabbar by repeating the fear litany before the pain begins. He has been carefully primed by Mohiam's suggestions to interpret the strange prickly sensation and growing warmth in his hand—his now-invisible *missing* hand—as pain.

One effect of the gom jabbar held at his neck is to force Paul into eye contact with Mohiam. She has locked him in a stare and may well be using hypnotic powers: "His world emptied of everything except that hand immersed in agony, the ancient face inches away staring at him" (*Dune* 9). The power of hypnotic suggestions to maximize or minimize pain emotions has been well established. In the final chapter, when they meet again, we are told "the old woman locked eyes with him," and Paul (now Duke Paul Maud'Dib) reacted angrily: "Try your tricks on me, old witch" (*Dune* 477). This response suggests that he now knows she used tricks—hypnotic suggestion—on him before, during their only other encounter.

The application of hypnotic suggestion to increase or decrease pain perception has more hard science to support it than the Pentagon's very speculative efforts to attain long-distance crowd control by means of ray guns. A group of Montreal-based researchers led by Catherine Bushnell has done a series of studies using hypnotic suggestion as a tool for demonstrating the role of cognition in pain perception—demonstrating in the process that pain can be enhanced or diminished by hypnosis.

The experience with Mohiam's box and gom jabber is clearly crucial to Paul's development. He has been carefully trained in combat by Duncan Idaho, but the testing by the Reverend Mother is the first threat to his survival, and it begins his transformation. Mohiam says the purpose of the test is to determine whether Paul is human or animal, but in fact the pain and the threat to his survival are the vehicles to awaken Paul's latent superhuman powers as Kwisatz Haderach. Paul emerges from the ordeal with a sense of "terrible purpose" (*Dune* 11)—an awareness of some drive or objective that awaits in his future.

His sense of terrible purpose resurfaces as he and his mother must flee into the desert at the same instant that his prescience expands far beyond the prophetic dreams of his boyhood: "He sensed that his new awareness was only a beginning, that it was growing. The sense of terrible purpose he'd first experienced in his ordeal with the Reverend Mother Gaius Helen Mohiam pervaded him. His right hand—the hand of remembered pain—tingled and throbbed" (*Dune* 188–189).

CAROL HART, Ph.D., is a freelance health and science writer based in Narberth, PA, just outside of Philadelphia. She is the author of *Good Food Tastes Good: An Argument for Trusting Your Senses and Ignoring the Nutritionists* (forthcoming, SpringStreet Books) and *Secrets of Serotonin* (St. Martin's Press, 1996, with a revised and expanded second edition forthcoming in early 2008).

References

Bromm, B. and R. Treede. "Human Cerebral Potentials Evoked by CO_2 Laser Stimuli Causing Pain." *Experimental Brain Research* 67 (1987): 153–162.

Hambling, D. "Details of U.S. Microwave-weapon Tests Revealed." *New Scientist* (online version), July 22, 2005.

Hambling, D. "Maximum Pain Is Aim of New U.S. Weapon." *New Scientist* (online version) March 2, 2005.

Herbert, F. *Dune*. New York: Berkley Books, 1977.

Hofbauer, R. K., P. Rainville, G. H. Duncan, and M. C. Bushnell. "Cortical Representation of the Sensory Dimension of Pain." *Journal of Neurophysiology* 86 (2001): 402–411.

Keefe, F. J., R. B. Fillingim, and D. A. Williams. "Behavioral Assessment of Pain: Nonverbal Measures in Animals and Humans." *ILAR Journal* 1991; 33.

Price, D. D. "Psychological and Neural Mechanisms of the Affective Dimension of Pain." *Science* 288 (2000): 1769–1772.

Rainville, P., G. H. Duncan, D. D. Price, B. Carrier, and M. C. Bushnell. "Pain Affect Encoded in Human Anterior Cingulate but not Somatosensory Cortex." *Science* 277 (1997): 968–971.

NAVIGATORS AND THE SPACING GUILD

John C. Smith

*Mind-expanding drugs vs. study and hard work—that's the fun-
damental difference between navigation in the Duniverse and
modern day NASA. JPL Navigator John C. Smith compares the
modern rocket science of navigating spacecraft to the planets to
the tasks of the Guild Steersmen with their spice-induced prescient ability to see future outcomes. Hey those navigator guys
would be great in Vegas—if you could get a large tank containing a hideous mutant floating in orange spice gas past the security cams.*

IMAGINE SPACECRAFT NAVIGATORS, stoned out of their
minds, pitching their laptops and workstations into the recycle
bin, to be replaced by lava lamps and incense burners in order to
guide today's spacecraft across the Solar System. NASA no longer exists. Instead, the navigators are under the employ of a huge for-
profit business that controls all space transportation throughout the
world. This business wields tremendous political power, to the point
where they can even replace heads of state at their convenience. Regarding those navigator positions—because of their inherent inferior-
ity in math and science, women need not apply!

Welcome to the Dune universe, where the Spacing Guild's prescient,
spice-saturated Steersmen navigate huge Holtzman-drive-powered
Heighliner ships safely through folded space—the only means of interstellar transport throughout the known galaxy. Guild ships do not land
but rather provide transportation from orbit about one planet to orbit about another. Since thinking machines have long since been outlawed, the spice melange gives navigators the prescient ability to see
the future of all possible outcomes so that they can choose safe routes

for their vessels. As the Guild is the sole employer of the navigators, it maintains a monopoly on interstellar travel, giving the Guild tremendous power and political clout.

The importance of the Guild in the Dune universe was succinctly stated by the Baron Vladimir Harkonnen, "we've a three-point civilization: the Imperial Household balanced against the Federated Great Houses of the Landsraad, and between them, the Guild with its damnable monopoly on interstellar transport" (*Dune* 17). The Imperial dating system is even tied to the start of the Guild's monopoly on space transport (B.G. is used for "Before Guild"). In the words of Reverend Mother Gaius Helen Mohiam, the Butlerian Jihad "forced human minds to develop" and therefore schools were started to develop human talents. There were "two chief survivors of those ancient schools: the Bene Gesserit and the Spacing Guild. The Guild, so we think, emphasizes almost pure mathematics" (*Dune* 8). Intricate mathematical knowledge combined with their prescient abilities enables the Guild's navigators to pilot their ships from one planet to another. Interestingly, the navigators (with one exception) are all male; the Bene Gesserit are all female.

In the Dune universe, navigators are viewed with awe and mystery. *Star Trek*'s vision of navigators is typically less flattering—such officers occupy the lower rung of the command structure, positions appropriate for ensigns and even acting ensigns. In *Farscape*, Pilot, a multi-limbed being with tendrils bonding it to the living ship Moya, acts as navigator, companion, and liaison. R2-D2 even served as a navigation droid for Anakin's starfighter during the attack on the Trade Federation Droid Control Ship. In science fiction, navigation usually only becomes important to the story line when the ship gets lost. For example, in the original *Trek*, the Medusan Ambassador Kollos is called upon to save the day when the *Enterprise* gets seriously lost beyond the edge of our galaxy. In *Babylon 5*, an entire episode is devoted to the rescue of the Explorer class ship *Cortez* when it becomes lost in hyperspace.

Navigators are often portrayed as either bizarre creatures or, as in the case of Ambassador Kollos, ugly enough to drive even a Vulcan insane. At least Herbert appropriately gave the navigators superhuman abilities and status in *Dune*. Of course, I could be biased since my career involves the navigation of unmanned interplanetary spacecraft. Upon reflection, perhaps the archetype of navigator as hideous mutant explains my social life.

The Space Business

Since the Spacing Guild maintained a monopoly on interstellar travel—as well as Imperial banking—the Guild was able to charge exorbitant fees for their services. Complaining about the high cost of military transport, the Barron Harkonnen commented to Rabban, "If you squeeze Arrakis for every cent it can give us for sixty years, you'll just barely repay us" (*Dune* 182). Similarly, the earthbound airlines of today, not unlike the maritime fleets of centuries past, are the only businesses that enable people to travel great distances in a short time. No monopoly exists, however, since any company—given adequate resources—can build or buy an airplane and train pilots to fly it.

Space travel is orders of magnitude more difficult. Until the launch of China's first taikonauts in 2003, the U.S. and former Soviet Union controlled all manned access to space. This chokehold is quickly changing, however, as private companies such as Virgin Galactic and SpaceX are entering the picture—motivated by the potential profits of space tourism. NASA is also funding the for-profit private firms SpaceX and Rocketplane Kistler to demonstrate the capability to transfer cargo to the International Space Station as early as 2010. Similarly, robotic missions to the planets were the sole domain of the U.S. and former Soviet Union until the 1990s. Now the European Space Agency and Japan have joined the ranks of this exclusive club and other newcomers are possible in the coming decades. In our lifetimes, however, the majority of interplanetary transportation endeavors are likely to remain in the hands of a select few nations due to the cost and complexity involved. Interplanetary travel is unlikely to be a profitable business anytime soon.

Role of the Navigator

In Dune, it's neither the expense nor the infrastructure that gives the Guild their monopoly: it's that they are the only ones with access to navigators, without whom interstellar travel would be very risky. From the

Legends of Dune series, prior to the genesis of the spice-mutated navigators, only nine out of ten Heighliners made it to their final destination. Navigation is the science of getting spacecraft (or anything else) from place to place, and generally involves two elements. First, a navigator needs some method of determining the current position of his craft with respect to some known reference, just as the ancient mariner triangulated his position and then plotted it on a latitude/longitude map. The next element of navigation is the prediction, based upon the present location and speed, of the future path the craft will follow, and then steering the desired course. The person responsible for these tasks is the navigator, also referred to in Dune as a Guildsman or Steersman.

To Saturn . . . and Beyond

In Frank Herbert's Dune trilogy, few details are provided as to how the navigators accomplish their tasks. Techniques used for interplanetary navigation today illustrate some of the problems the navigators of Dune would face and provide insight into perhaps why Herbert endowed them with superhuman prescient abilities. Although manned interstellar travel is currently well beyond our reach, robotic interplanetary travel is a daily reality. In fact, four unmanned spacecraft (Voyagers 1 and 2 and Pioneers 10 and 11) are already headed for interstellar space, having long ago completed their primary missions. The most distant human-made object is the Voyager 1 spacecraft, launched in 1977, which is now more than 15 billion kilometers from the Sun (more than twice the average distance of Pluto). After completing its primary mission of exploring the Jupiter and Saturn systems, Voyager 1 should reach the boundary of interstellar space within the next decade. It's estimated, however, that it will take 40,000 years for the craft to reach the vicinity of the next star on its path, located in the constellation Camelopardalis. I'll opt for the Guild Heighliner any day.

On Earth, the chief responsibility for navigating spacecraft to the planets is the purview of NASA's center for unmanned space exploration, the Jet Propulsion Laboratory (JPL) in Pasadena, California. The steps in navigating a modern spacecraft are conceptually similar to navigating an airplane. For example, before boarding a flight from your home to Las Vegas, someone had to construct a flight plan, i.e., plan the nominal path and schedule, predict how much fuel will be required, and determine which traffic-control ground stations would be used to monitor its progress. The pilot/navigator continually determines where the plane is using various measurements, such as ground tracking, GPS, and visual landmark identification. Once the plane's location is determined, the pilot updates the prediction of the plane's future path (to ensure an on-time arrival at the destination) and must occasionally adjust the direction of the plane to keep it on course since winds and other factors aren't perfectly predictable.

Engineers at JPL navigate spacecraft to much more exotic and distant locations than Las Vegas. Armed with a firm knowledge of how the forces of nature—particularly gravity—govern planetary and spacecraft motion, engineers first design a baseline path for the spacecraft to follow from the spacecraft's current position to its intended destination, such as from Earth to Saturn. Once the craft is launched, JPL navigators periodically update its present location and prediction of its future path as often as every few hours during critical activities or as little as every few weeks for spacecraft well past their prime but still phoning home. Small rocket burns (maneuvers) are used to maintain the spacecraft near its baseline trajectory, but this course is never exactly followed any more than an airplane exactly flies its flight plan. However, such perfection is neither required to achieve the goals of a space mission nor to fly you to the slot machines by happy hour.

Can You Track Me Now?

Earthbound navigators must pilot interplanetary missions remotely since human exploration of the planets is, most likely, decades away. Today's planetary spacecraft are controlled via radio transmissions from antennas from one the of three major Earth-based antenna complexes composing the Deep Space Network (DSN). The three antenna complexes are spaced nearly equidistant in longitude such that any spacecraft will always be in view of at least one of the following sites: Goldstone, California; Madrid, Spain; and Canberra, Australia. The same radio signal used to command the spacecraft is used to determine the spacecraft's state—its position and velocity—to incredible accuracy. For example, even though Saturn is nearly a billion kilometers away, the state of the Cassini spacecraft currently orbiting Saturn is typically determined to within about 2 kilometers in position. Such accuracy is required since future plans call for the Cassini craft to pass within 25 kilometers of the surface of Saturn's geyser spouting moon Enceladus. Other measurements, such as optical imagery of planets and/or star fields from a craft's camera, may also be used to further refine a spacecraft's position.

Even more impressive, when a radio link is established between an antenna on the Earth and the spacecraft, measurement of what is known as the *Doppler shift* of the radio signal provides an extremely accurate measurement of the spacecraft's velocity relative to the Earth along this line of sight—an accuracy typically to within 1/100 of a meter per second. The measured Doppler shifts are not direct measurements of the spacecraft's state, they're just measurements like multiple compass bearings taken while hiking a trail. A multistep process is required to process these measurements to determine the state of the spacecraft at each moment it is in direct communication with a DSN station.

(Image courtesy NASA/JPL/Caltech)

The radio waves used to both communicate with the spacecraft and to determine its state travel across space at the speed of light. The round-trip time for the radio signal varies over time since all bodies are in motion (even the stars). For a Mars mission, the round-trip light time can vary between eight and forty minutes. For a Saturn mission, 200 minutes is typical. This communication delay time is often ignored in science fiction, but is a practical complication for modern-day navigators.

Dude, Where Is My Spacecraft?

The first step in interplanetary navigation is to determine the spacecraft's position and velocity, its *state*, at a given moment in time. You can't know where you're headed, if you don't know where you started. Guild Navigators would also have to have some idea of the initial location of their ships, since the safest path to their destination must have a starting point. The more accurately the initial state is known, the more accurately the craft's future course can be predicted.

A spacecraft's state is specified with respect to some known reference frame. Just as islands and other familiar landmarks served as a reference frame for ancient mariners, pulsars might be used for interstellar travel. The plaques on both the Pioneer and Voyager spacecraft, both bound for interstellar space, depict the Earth's location using pulsars as references. Radial lines originating from the Earth indicate directions to fourteen pulsars. These pulsars can be identified by their rotational periods depicted graphically using time units based on the electron state transition of the hydrogen atom (the most common element in the cosmos). References used for current-day interplanetary navigation are the Earth's North Pole and the intersection of the Earth's equatorial and orbital planes on January 1, 2000.

The spacecraft's state is determined throughout the period it is in radio communication with the Earth since that is the only time period for which new measurements are available. The first step is to make a reasonable guess of its current path. Returning to the airplane analogy, if two hours ago the plane you were navigating flew over Chicago heading west at 800 kilometers per hour, a reasonable guess of your new position would be 1,600 kilometers west of Chicago. To get a

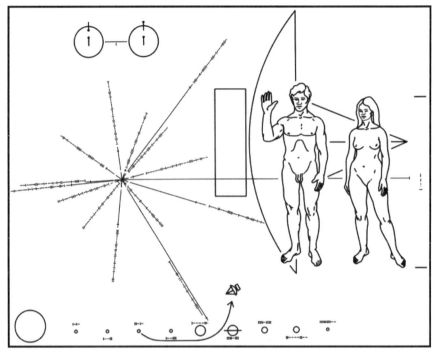

(IMAGE COURTESY NASA/JPL/CALTECH)

higher precision estimate of the plane's new position, you would need to take new measurements.

To produce a guess of the spacecraft's current path, its last known state, perhaps from yesterday, is predicted forward in time using mathematical models of the many forces acting on it. The longer you try to propagate your past knowledge into the future, the greater the uncertainty in the prediction—think of how quickly a weather forecast degrades in accuracy. Gravity is usually the dominant force acting on a spacecraft, but many other forces must be modeled in modern navigation to achieve the high accuracies required by today's missions. Such forces include the non-symmetrical distribution of mass in planets and moons (their lumpiness), radiation pressure from the Sun (the solar wind), drag if an atmosphere is nearby, rocket firings by the spacecraft, etc. The cumulative effects of these forces are modeled using computers to obtain a predicted path and these computers are definitely *not* thinking machines. The state determination process is not an exact science since we lack the knowledge to perfectly model the physical universe, just as we can't perfectly model Earth's atmosphere to make exact weather predictions. Interstellar travel would have even more non-gravitational forces that would be difficult to model/predict.

High-precision predictions of the state of the Earth (and all other natural bodies in the Solar System) already exist. Given a prediction of the state of the Earth and the spacecraft, the Doppler shift in the radio signal is also predicted and then compared to the actual Doppler shift measured while the spacecraft is in communication with the Earth. Differences between the predicted and measured Doppler shifts occur since we can't exactly predict all the forces acting on the spacecraft. Back in cubicles on Earth, navigators tweak mathematical model parameters until the differences between the predicted and measured Doppler shifts are minimized.

Using these finely tuned models, a best-estimate prediction of the spacecraft path, its *trajectory*, is generated. This trajectory yields the best estimates of the spacecraft position and velocity throughout the entire time the craft is in communication with the Earth. Spacecraft, planetary, and satellite trajectories created at JPL and other NASA centers are stored for worldwide distribution in a special format known as a SPICE kernel. You might say that modern-day navigators create SPICE rather than consume it!

Where Are You Going?

Spacecraft trajectories are usually propagated well beyond the latest tracking pass in order to assess spacecraft safety—such as ensuring the spacecraft doesn't impact a planet or moon—and determine whether the craft is on course for its destination. There is always uncertainty in the best estimate of the spacecraft state and in any computational propagation of that state into the future. In other words, any computer prediction of a spacecraft's state will be less accurate the more distant it is projected into the future. This uncertainty of an object's position in an orbit always grows with time. Consider that the next time someone (who is probably selling a doomsday book and sowing fear to generate publicity) makes a specific prediction that an asteroid will wipe out the Earth on a given date. It is simply not possible to accurately predict Earth-threatening asteroid trajectories more than a few years into the future (which doesn't mean, however, that we should ignore the possibility).

The inability to exactly model the forces acting the spacecraft contributes not only to the uncertainty in determining the current spacecraft state but also to uncertainty in the predicted state. An example that has received a lot of publicity is the "Pioneer anomaly." Both Pioneer 10 and 11 are among the most distant objects tracked by earthbound navigators. Although the states of the Pioneers have been accurately determined numerous times, predictions of their future courses have been inexplicably wide of the mark even when considering the effects of known uncertainties. Many have speculated that this discrepancy demonstrates that our knowledge of the laws of physics is wrong, and in particular, that changes to the law of gravity may be required to explain the discrepancy. However, the most likely, though yet unproven explanation, is that improper modeling of the non-gravitational force due to radiation pressure from the radioactive power source onboard the Pioneers is to blame. Guild Navigators never need be concerned about errors in their predictions since they can see into the future—they are concerned with the effect, not the cause.

Dune Navigators (Slugs on Parade)

In Herbert's first book, Guild agents make a rare appearance in the climatic finale as "the two fat ones dressed in gray over there"—a rather unremarkable description other than a mention of their eyes appearing "a total blue so dark as to be almost black" from spice consumption (*Dune* 367; 359). In *Dune Messiah*, one of the main characters is a navigator named Edric, who lives in a tank that simulates weightlessness surrounded by orange spice gas. The Steersman is described as "an elongated figure, vaguely humanoid with finned feet and hugely fanned membranous hands—a fish in a strange sea" of orange spice gas (*Dune Messiah* 9). Edric clearly displays a greater degree of mutation than the navigators in the first Dune novel.

In the David Lynch feature film, a "third-stage" Guild Navigator is portrayed as an embryonic-looking sluglike creature floating in a tank of spice gas (a portrayal which, again, does nothing beneficial for my social life). An interesting gaffe in the movie portrayal is that the navigator lacked the blue eyes caused by spice consumption. Obviously, the navigator was a strange being, but as an earthbound interplanetary navigator myself, I've seen some strange-looking people at my work place in Los Angeles.

Prescience

The spice-mutated Guild navigators use prescience, the ability to see all future possibilities, to guide the Heighliner ships through foldspace. Since thinking machines in the Dune universe were banned after the Butlerian Jihad, Herbert recognized that space navigation would be a difficult task for mere humans and thus endowed his mutants with superhuman abilities. The Dune navigator can see the future outcome of each possible trajectory the ship might take, thus choosing a course that safely reaches the desired destination. In *Dune*, Paul refers to navigators as "men who can quest ahead through time to find the safest course for the fastest Heighliners" (*Dune* 345). The spice-induced prescient knowledge allows the Dune navigator to circumvent the incredibly complex mathematics and physics problems associated with interstellar navigation, and their ability to see all possible paths is reminiscent of the quantum mechanics many-worlds hypothesis.

Eight years before the publication of *Dune*, physicist Hugh Everett III proposed a radical new way of interpreting some of the bizarre behavior observed in quantum mechanics. Everett proposed that everything that can happen *does* happen. Each possible action spawns a new universe. The number of resulting multiverses, or parallel universes as they are sometimes called (science fiction staples often portrayed in *Stargate* and *Star Trek*), increases geometrically over time, but no communication between the universes is possible. The prescient ability of the Dune navigator to choose a safe path seems to imply that they can see the path of their ships in many, if not all, multiverses, defying the laws of physics as we understand them (or hypothesize them to be), but that's why prescience is a superhuman ability.

It's unclear whether the navigator sees *all* possible paths or just searches until the *first* safe path is identified. The latter "first fit" option would be more computationally efficient. If there are infinite paths, then there are infinite safe paths, so there is nothing to be gained in choosing one safe path over another. Even very fast computers can take an ungodly long time to find the "best fit" solution to any problem, whereas finding one acceptable solution can be quite fast. In *Dune*, a first fit search is one possible interpretation of Paul's dread of "the idea of living out his life in the mind-groping-ahead-through-possible-futures that guided hurtling spaceships" (*Dune* 150). Navigators are not omniscient but that spice sure does come in handy.

Herbert's gift of prescience permits the Guild Navigators to ignore sources of error in determination of the spacecraft's current state and predicted path since their ability allows them to see the cumulative results of all the forces the physical universe can throw their way. The results of uncertainty and mis-modeling are already incorporated in the many trajectories from which they can pick and choose. In essence, they can choose a blue sky and sunshiny location 100 percent of the time without the need to model the weather. Modern-day navigators aren't that lucky. For example, the location and gravitational attraction of the planets and moons are never known perfectly, the Sun has its own solar weather cycle, atmospheric drag can be as unpredictable as the weather, and/or the spacecraft itself might not perform maneuvers in exactly the manner predicted. There are also sources of what's referred to as "noise" in the measured radio signal due to such things as attenuation by Earth's atmosphere or something less obvious like deformation of the huge 70-meter DSN antenna due to wind or bearing friction. These and other sources of uncertainty must be tweaked by JPL engineers for every batch of measurements taken during communication with Earth to get the best fit spacecraft orbit—a time-consuming and sometimes tedious task.

Do Men Make Better Navigators than Women?

In *Dune*, two ancient schools survive from the times of the Butlerian Jihad—the Spacing Guild and the Bene Gesserit. These schools were designed to develop human talents to eliminate dependence upon "thinking machines." With one exception, the Guild is male and "emphasizes almost pure mathematics" (*Dune* 13, 394, 399). The Bene Gesserit is female and emphasizes mental and physical training and their agenda features the breeding of a super being, the Kwisatz Haderach (a male). The split between Guild and the Bene Gesserit is along left brain/right brain lines. Herbert's gender assignments touch on a very contentious issue. In general, are men inherently better suited to be navigators than woman?

There is no dispute that women are underrepresented in math and science careers. The issue is why. Explanations usually boil down to extreme nature, extreme nurture, or somewhere in between. The "nature" position would be that men are biologically better suited for science and math. As one Harvard professor put it, "only a madman would take that stance." The "nurture" position is that males and females are biologically indistinguishable and that all gender differences are a result of socialization and bias. The majority of opinions fall in between these two extremes.

An example of the volatility of this issue can be seen from the uproar that resulted from a speech given by Harvard University's president, Lawrence Summers, in 2005 which leaned toward the nature explanation. He was subsequently forced to resign. The amount of literature devoted to this subject is staggering and often contradictory. I doubt even a Mentat could make sense of it. There appears to be general agreement that there are indeed some differences in specific abilities between the sexes, but whether such differences add up to an overall advantage is another matter.

Any comparison must be statistical in nature since there is no question that there are many women who are better at math and science than men and visa versa. In Frank Herbert's *God Emperor of Dune*, it's even revealed that the first Guild ship was designed by a woman (Norma Cenva). In 2004, Brian Herbert and Kevin Anderson expanded her role to that of the first spice-mutated navigator, but personally I believe this was a bit of political correctness. Most statistical studies conclude that, on average, men and women show no difference in general intelligence. However, a common argument by the nature camp is that while the means of math/science ability distributions are similar between the sexes, the tails extend further for men. In other words, men contribute more prodigies and idiots. Whether this is really true, like most assertions, is hotly contested, but when's the last time you heard of a woman earning a Darwin Award "commemorating those individuals who ensure the long-term survival of our species by removing themselves from the gene pool in a sublimely idiotic fashion"?

Common differences ascribed to men are that they are more career oriented, have a greater interest in things versus people, and are greater risk takers. Women are often ascribed to having better verbal ability, emotional perception, and recall. The nurture camps are quick to correctly point out that such behaviors can be explained by social environment. Separating out the effects of environment from biology may be an insurmountable task for humans since there will never be a control group of people who can be considered to have been raised in a neutral environment. Studies have shown that even parents who think they are not reinforcing gender stereotypes often do so unconsciously.

A rare area of common agreement seems to be that men perform better than women in a variety of spatial visualization and navigational tasks such as mentally rotating an object in three dimensions or predicting the trajectory of a missile. These abilities would be useful to the Guild Navigator's tasks but navigation requires much more than just these skills. Women get better grades than men in school at virtually every level of education, including in science and mathematics, and half the math degrees awarded by colleges are to women.

A 2007 *Scientific American* article makes the case that the most important factor may be the level of exposure to various sex hormones early in life. Testes produce male hormones that appear to "alter brain function permanently during a critical period in prenatal or early postnatal development." A small but interesting study by Mark Brosnan of the University of Bath found that people whose ring fingers are longer than their index fingers usually scored higher on the math portion of the SAT test than the verbal portion. Exposure to testosterone in the womb is said to make the ring finger longer and also promote development of areas of the brain associated with spatial and mathematical abilities. A Duke University researcher has found that male and female rodents solve problems differently and that male rats navigate mazes more efficiently.

Physical differences between the sexes in the size and makeup of the brain are often discussed by the nature contingent. Male brains are 10 percent larger than female brains and some argue that gives men an edge. However, by that logic, elephants should be geniuses. Also, when comparing men and woman of the same weight, brain size is quite similar. From birth, the *corpus callosum*—nerve tissue that connects the right and left sides of the brain—appears to be larger and more extensive in women, causing many to speculate that this leads to superior language and communication skills. According to David Geary, a professor of psychological sciences at the University of Missouri, "females use language more when they compete. They gossip, manipulate information." Sounds like a Bene Gesserit to me!

There are sex differences, but they don't necessarily add up to an overall advantage. In science, we need to keep an open mind and not prejudge the outcome based on our belief systems. We may simply not have the scientific means at our disposal to answer this question in our lifetimes. My own personal experience as an interplanetary navigator working on NASA missions is that men have always dominated the field but that this imbalance is slowly changing. For example, in the early 1980s, it was unusual to see more than one woman in the room. Today, JPL's flagship mission—Cassini/Huygens—is nearly one-quarter female. So either woman's brains have evolved more in the last twenty years than in the last 50,000 years, or most, if not all, of the workplace gender disparity is can be traced to environmental and social differences.

Perspective

Today, highly specialized engineers perform interplanetary navigation to impressive precision, but there is nothing magical about it. The tasks of determining where a spacecraft is and where it is going will forever be key undertakings. Fast computers may do the number crunching, but the adage "garbage in, garbage out" is in full force. The hardest navigation decisions today are still made by human beings. Navigating interstellar flight may appear daunting to us now, but prior to the publication of Frank Herbert's *Dune* in 1965, interplanetary navigation may have seemed comparably difficult.

During the time period *Dune* was written, humanity's exploration of the moon and planets was in its infancy. The first successful flyby ever of another planet was NASA's Mariner 2 craft, which encountered Venus in late 1962. In 1965, Mariner 4 became the first craft successfully navigated to encounter Mars. Prior to 1965, these were the only two missions out of twenty that were launched to successfully complete the journey to their interplanetary destinations. It's therefore not surprising that Herbert gave navigators superhuman prescient abilities to tackle the even more daunting problem of interstellar flight. Perhaps if the Dune series originated today, the author would not have given his navigators the gift of prescience but instead the gifts of a Mentat. It also seems fitting that the first two successful missions to the planets were named Mariner, since a mariner is a person who navigates or assists in navigating a ship.[1]

JOHN C. SMITH has a master's in celestial mechanics (the motion of heavenly bodies) and has spent more than twenty years at NASA's Jet Propulsion Laboratory in Pasadena, CA, designing missions to explore the planets. He has worked on successful missions to Venus, Earth, and Mars and has been part of the Cassini/Huygens mission to Saturn and Titan for sixteen years. He is the designer of the four-year "tour" of the Saturn system which began in 2004 and recently contributed to the design of a two-year extended mission through 2010.

[1] "Dune" entry from Merriam-Webster online dictionary, 2007 <http://www.merriam-webster.com/cgi-bin/dictionary>.

MEMORY (AND THE TLEILAXU) MAKES THE MAN

Csilla Csori

Are we the sum of our memories? If so, given the fragile nature of Human memory—how easily memories can be lost, altered, or even created—what are we really? If that's not angst-inducing enough, what are you, then, if you're a clone made by the Tleilaxu, a ghola, and your memories—which aren't really even your memories—are all implanted? Csilla Csori explores these very issues.

SINCE MANKIND FIRST SAT AROUND A FIRE telling scary stories, we've had tales of people coming back from the dead in one form or another. Modern tellings substitute science for the supernatural, but these stories continue to capture our imagination and leave us wondering, is it possible? Would such a person be the mindless zombie of countless horror movies, or a resurrected Lazarus?

These questions will not always be academic. People used to be considered dead when their hearts stopped, but not anymore. Every day, people whose hearts have stopped beating for brief periods of time—sometimes their hearts are even stopped on purpose during surgery—are brought back to life, and they are essentially the same person as before. The next logical step is to consider what would happen if we could overcome brain death. Would the revived person be the same, have the same personality, and retain all of their memories?

In *Dune Messiah*, we are first introduced to the ghola as a body resurrected through medical means. Our ghola, Duncan Idaho, has a consciousness of his own but no memory of his former life or his death.

He is most definitely not a mindless zombie, and, in fact, is resurrected with the capability to learn new skills which he didn't possess in his previous life. He is a person, but is he the same person, the same Idaho who lived and died before? Not at first. Although certain voices and places seem familiar, he doesn't regain any actual memories until one traumatic event unlocks the past, and all of the memories from his former life come flooding back at once. If the brain is like a computer, then it is as if his memories are stored in a hidden file system to which he does not have access. Once he acquires the key to unlocking that system, all of the files (or memories) within are opened to him at once, and he knows himself as Idaho.

The news is full of stories in which authorities confiscate a suspect's computer and recover a hoard of incriminating files which the suspect had deleted. If a computer can retain deleted files, what about a human brain? Should this analogy of recovering information from a damaged brain give us hope for the Terri Schiavos of the world? If we could repair and regrow brain cells, would her personality still be in there somewhere, fully intact and just needing the right key to unlock it? To answer these questions, we need to examine how a brain stores and retrieves memories, and how this process compares with computer memory.

Like a computer, your brain has storage systems for short-term and long-term memory, and a central processing unit, known as the *hippocampus*, which connects the two. Your hippocampus filters data— determining what is relevant—from short-term memory to long-term memory. However, the manner in which data is stored and then accessed later is different, and that is where the key to recovering lost memories lies.

Most computer users are familiar with hierarchical file systems, which are basically made up of a main directory (or folder) containing files and subdirectories. The subdirectories can, in turn, contain both files and additional subdirectories. Users navigate up or down the hierarchical structure to locate a specific file in a specific directory. You might expect that all of the files in a particular directory are stored next to each other in the computer's memory, but this is not always the case. In fact, a single file may be broken into fragments and stored in several locations. This fragmentation occurs when files are edited and increase or decrease in size. On a brand-new disk (or other chunk of storage), the computer's operating system starts at one part, writing data in an orderly fashion, and if the data never changes, it continues until the disk is full. But data files are almost never static; users are constantly adding on to files, deleting entire files, and otherwise changing the amount of memory needed to store a subdirectory or a particular file. As data is deleted, chunks of memory become available for new information, making holes in the nice, orderly system. When a file increases in size, if there is not enough memory in the original location to store the entire file, then the computer will look for an additional chunk of memory to store the second piece of the file. This process can be repeated many times, and a single file may end up stored as several pieces spread out over the disk.

The exact method a computer uses to keep track of all the pieces differs between operating systems, but it basically uses some kind of master reference table. When a user deletes a file, the actual data is not erased—only the entry in the reference table gets deleted. This tells the computer that the particular chunk of memory on which that file is stored is now available for writing new data. But the old data will sit there until it is overwritten, so that is why it is possible to recover deleted files from a computer.

Does the brain work in an analogous way, allowing us to recover lost memories? Your brain also stores pieces of memories in different locations; but, unlike a computer, it does not store information sequentially. Different types of sensory signals, such as sight, sound, and taste, are processed in different regions of your cortex and routed to your hippocampus. After filtering, the hippocampus sends these bits of information back to their respective regions and creates neural links between them. These links are strengthened by repetition (for example, by repeating a list) and by emotional factors such as the personal relevance of the information. Your hippocampus keeps track of all of the links and associations, indexing and cross-linking with similar information. It seems similar to the master reference table in a computer operating system, but it is much more complex. Even though a computer may break a file into several pieces for storage, it still considers a file to be one discrete unit. The computer has no way of examining file content and determining that the letter you wrote to Grandma last week is in any way connected with the photo of her hugging you as a child. Your brain's reference system, on the other hand, is able to cross-link information from memories that are widely separated in time and location, and makes connections based on everything from strong emotions to mundane details.

This interweaving of memories strengthens associations, but it can also muddle memory recall and make it unreliable. When you recall the memory of an event, you are not opening a single file containing all of the data. Rather, you are dynamically reconstructing the memory from its various components. The process is associative, so one thing, like a particular song or smell, can trigger an associated piece of the memory, which triggers another, and so on. The ease and accuracy of your recall depends on the number and strength of the neural links, which, in turn, are dependent on such factors as how long ago the event occurred, when you last remembered it, and whether it is similar to other events in your memory. In the process, pieces of memories can get confused and mixed in with one another. For example, a married couple who has had several arguments over money may mix up what was said during which argument when trying to recall one particular confrontation. If they later have to testify in court as to what was said, they may give different accounts and yet each will believe they are telling the truth. In addition, the process of memory reconstruction is further clouded by current emotions and motivations. So, unlike a computer file, which is the same each time you open it, your memory of an event will differ at different times in your life.

Consider again our ghola, who has no memory of his former life. If those memories are still stored in his brain, how might they be accessed? Amnesia is often temporary, with people gradually recalling some or all of their missing memories. Our ghola's brain has been repaired, so there is no physical damage preventing access. If the neural links are intact, then it should be as simple as placing him in an environment which will trigger the old memories. It is unlikely that everything would return at once. A familiar face or voice would bring back a flood of associations, and, over time, the entirety of his memories should return. Of course, he is not exactly the same person, especially after the trauma of remembering his own death—but he is, for all intents and purposes, Idaho.

However, it's not that easy, because our ghola's memories are locked away in that hidden file system. In searching for a physical cause for the block, you might suppose that there is something in his hippocampus, or CPU, that is preventing access, but it is not that straightforward. Once the associations between neurons—the neural links—reach a certain strength, they become independent of the hippocampus, and the neurons can trigger each other directly. So his oldest, strongest, and most well-connected memories are not controlled by the hippocampus at all. In fact, damage to the hippocampus has the opposite effect on memory than what our ghola is experiencing. Rather than causing retrograde amnesia—the inability to recall past events—a damaged hippocampus causes anterograde amnesia—the inability to acquire new memories. Without the hippocampus, short-term memories can never be translated into long-term memories, and they are lost forever. Drew Barrymore's character, Lucy, in *50 First Dates* and Guy Pearce's character, Leonard, in *Memento* are two examples of people suffering from anterograde amnesia.

Therefore, there is no simple physical explanation for a total memory block in the presence of familiar surroundings. Due to the distributed, associative nature of memory, there is no central switch to turn on and off, no single access point which can be hidden or encrypted. Even in cases where a person suffers from severe retrograde amnesia due to lesions on the brain, such as in Alzheimer's disease, early childhood memories generally remain intact.

Perhaps our ghola's memory loss is not due to a physical cause, but a psychological one. The trauma of dying is surely something he would want to block out. Although rare, there have been cases where people suffered from amnesia after being the victim of a violent crime, but the amnesia was associated with a confused state and only lasted a short time. What remains, then, is the complex and controversial subject of repressed memories, a concept which is often associated with childhood abuse. Can a memory be forgotten, either intentionally or subconsciously, and then be remembered later? According to the American Psychological Association, both phenomena do occur, but the mechanism is not well understood. The accuracy of recovered memories is questionable; as the brain reconstructs those memories from their component parts, the person's emotions and intent influence the result. Memories are not perfect recordings of events, but rather, impressions colored by our emotional state both at the time the memory was formed and at the time it is remembered. To further confuse matters, it is possible to construct false memories of events that never occurred.

Even though the concept of repressed memory is possible, it does not offer a satisfying explanation for the total amnesia our ghola is experiencing. In recorded cases of repressed and recovered memories, the phenomenon was localized to those memories associated with the traumatic event. Our ghola might not remember the circumstances of his death, but he would not suppress the memories of his entire life. So a psychological cause for his type of memory loss is no more likely than a physical one.

We have looked at the question of access, of how memories are recalled, and whether they could be hidden from the conscious mind until triggered by a single event or whether memories would return in bits and pieces over time. But this assumes that the intact memories are in the brain to begin with. The next question is of storage, of whether old memories would remain in the brain at all. The answer depends on the type of ghola, since there are two distinct methods for creating them.

In the time when *Dune Messiah* is set, the process of creating a ghola requires the entire body of the original person. A ghola is literally a corpse brought back to life. The dead flesh of Idaho is placed in a tank where his damaged tissue is repaired, and a person emerges, alive and conscious. This person has no memory of his past, but since he has the same brain, he still has the neural connections (the file system of memories) created by all of the events of his life. Time is the greatest limiting factor, since neural links weaken with disuse. If a lengthy period passes before our ghola is exposed to memory triggers, some of his past may be lost. But his oldest and strongest memories will remain for a long time, and chances are good that he will regain at least some of his former identity.

However, by the time *God Emperor of Dune* takes place, technological advances have changed the process dramatically. The gholas of Idaho are not the same body repaired and resurrected again and again. They are grown from mere cells of the original person, and there can be more than one of them alive at any given time. In other words, they are clones. This has completely different implications for the possibility of memory retention, because it requires the transfer of memories from one body to another. Like any clone, the adult gholas of Idaho are created using DNA as the means of coding information into the copy. What we know about the way memories are stored and retrieved in the brain involves neurological and chemical processes. There is no research to indicate that DNA stores specific memories, such as the events in a person's life. As our ghola grows in his tank, his DNA dictates the basic structure of his brain, but it does not stimulate the neural links which are key to the creation of memories. When he emerges, even though he is physically an adult, he is essentially a newborn person. Unfortunately, our Idaho clone has no inherent memories of the original Idaho's life.

What about transferring memories, downloading them from the original into a copy? Preserving the original brain indefinitely poses a problem, so it is more practical to download memories into a permanent storage system, such as a computer disk or flash drive, and upload the information into our ghola as needed. This system requires a working interface between the computer, the hippocampus, and other parts of the brain; but once that is achieved, it is a matter of sending signals through the brain and recording the position and strength of electrical impulses. This gives us a snapshot of the physical structure of Idaho's brain at the time of his death.

If we re-create this physical structure in our ghola's brain, is it the equivalent of uploading Idaho's memories? More than a question of physical and biochemical requirements and limitations, the heart of this query asks what makes us who we are. If we can create physical clones of Idaho and give them all the same memories, experiences, and personality, then what makes any of them a unique individual? If the clones are perfect recreations, do terms like "original" and "copy" even have any meaning? The conclusion of these questions may have to wait until the first ghola emerges from his tank and speaks the answer.

Until cloning reaches that level of technology, our first type of ghola—the resurrected person—is the kind we will have to deal with. It is not just a subject for speculative fiction, but a topic for present-day discussion. As medical science advances, the moment when a person is beyond resuscitation gets pushed further and further back. Like Miracle Max in *The Princess Bride*, our doctors can determine if a person is just "mostly dead," and therefore partly alive. Machines can assist the heart and lungs to function until the body heals sufficiently to work on its own. Unfortunately, brain science has not yet advanced to the point where we can repair damaged brain cells, but that, too, is in our near future.

What will a real-life ghola, a person returned from brain death, be like? Will he remember any of his past, or will he be an entirely different person? In addition to impaired function, people who suffer from non-lethal brain damage often experience memory loss and even changes in personality. Repairing damaged cells would clearly return them to normal functioning, but what about memories? A cell which sustained only partial damage would retain some of its neural connections. A newly grown brain cell would not, but if it were connected to undamaged cells, the links from those healthy cells might be sufficient for the memory connection. Memory recovery would depend greatly on the extent of initial damage, but the distributed nature of memory works to our benefit here, as it's unlikely that all areas associated with any particular memory would have been damaged.

When medical technology provides us with a method for repairing and regrowing brain cells, the diagnosis of brain death may cease to exist. Just as people who suffer cardiac arrest today can have their hearts restarted, people who suffer severe brain damage may someday have their brains jump-started, or otherwise brought back online. For the person returned to life in this manner—our real-life ghola—this means that he has a chance of regaining at least parts of his memories, especially if he is in familiar surroundings which will trigger memory associations. And that's really all our ghola needs: a chance of recovery, the hope that his memories may trickle back and that he will regain some semblance of the person he was before.

CSILLA CSORI is a programmer/analyst at the San Diego Supercomputer Center. She works primarily on database and software development for business applications, and she also moonlights as a gremlin hunter for her colleagues when their computer programs start acting funny. Recently, she released version 5.1 of ProBook grant application software she authored for the University of California. It's one of those pesky projects that started small but took on a life of its own, and now, like *Doctor Who*'s Cybermen, keeps coming back to demand more upgrades. She gained an interest in brain function in college, where she earned extra cash by volunteering for cognitive experiments at the National Institutes of Health. In her spare time, she enjoys playing softball, kayaking, and any other excuse to be outdoors in San Diego's perfect weather.

References

American Psychological Association. "Questions and Answers about Memories and Childhood Abuse." *Learning and Memory.* Aug. 1995 http://www.apa.org/topics/memories.html

Dubuc, Bruno. "Memory and the Brain." *The Brain From Top to Bottom.* http://thebrain.mcgill.ca/

COSMIC ORIGAMI: FOLDED SPACE AND FTL IN THE DUNIVERSE

Kevin R. Grazier, Ph.D.

Faster than a speeding bullet—300,000 times faster—the speed of light in a vacuum is the speed limit of the Universe. It also tends to be a wet blanket for science fiction writers if strange distant planets are so distant as to be unreachable. Then there are all those nasty relativistic effects. If the action in our story takes place on different worlds, though, faster-than-light (FTL) travel is a requirement. How has this been handled in science fiction in general, and in specific within Dune?

I T HARDLY PUSHES CREDIBILITY to say that in the early 1960s and prior, the concept of traveling at superluminal—faster-than-light—speed was hardly part of popular culture. In 1965, however, millions of television viewers were given their first real indoctrination to the concept of getting somewhere in a *serious* hurry when they were presented with "Mr. Sulu, ahead warp factor five."

The updated 1990s version was a snappier, "Mr. Crusher, engage."

A new century yielded a different mindset. "Chevron seven, locked!"

In 2004 we got the minimalist version. "Jump!"

Today if a friend said to you, "We have ten minutes to get there, we need to drive at warp speed," even the most adamant of non-science fiction fans would understand the reference. Even a common phrase in business and the military, "I want it done yesterday," can, at some level, imply that a faster-than-light journey—at least from Albert Einstein's relativistic perspective—will be required at some point along the way.

Faster-than-light travel, FTL, has been a mainstay of science fiction for decades. Different novels/television series/movies have referenced different technologies which have, in turn, implied different methods of implementing FTL, but even before Captain Kirk gave his first command to take *Enterprise* to warp, Guild Steersmen navigated Heighliners through folded space with passengers, cargo, troops, and shipments of the spice melange from Arrakis. Unlike other works, where it is generally implied that complex (electronic) computation is a necessity for such travel, the Dune novels gave us a different approach. Guild Navigators, or Steersmen, empowered with the limited prescient abilities gained from ingesting spice, had the ability to navigate vast distances safely. "(Edric) ate the spice and breathed it and, no doubt, drank it, Scytale noted. Understandable, because the spice heightened a Steersman's prescience, gave him the power to guide a Guild Heighliner across space at translight speeds. With spice awareness he found that line of the ship's future which avoided peril" (*Dune* 5).

In the first novel, *Dune*, upon arriving at a new home on the planet Arrakis, Duke Leto Atreides stressed to his son Paul the necessity to develop an entirely new methodology for ensuring their tactical superiority. "Our supremacy on Caladan," the Duke said, "depended on sea and air power. Here, we must develop something I choose to call desert power" (*Dune* 58).

The planet Arrakis had but a single resource to supply to the Empire: the spice melange (unless, of course, we include religious fanaticism, but that came later). The politically powerful Guild Navigators required the spice to do their work:

> Not without reason was the spice often called "the secret coinage." Without melange, the Spacing Guild's Heighliners could not move. Melange precipitated the "navigation trance" by which a translight pathway could be "seen" before it was traveled. (*Children of Dune* 84)

Arrakis was the single source of spice, and since those who controlled the desert controlled the flow of spice, one could argue that the Guild's reliance on the drug represented what engineers today call a "single point failure mode." This weakness would be both exploited and addressed in various novels.

The Guild Navigators, in turn, with their unique navigation capability had developed "space power." Arrakis had but the spice to trade, and the Guild had a monopoly on FTL travel. Although this was more of a political/business form of power than a military one, it gave the Guild a leg up in nearly all of their negotiations. At a root level, much of the political intrigue of the Dune series, in fact, stems from the interplay of space power and desert power. What would it take in practical terms for the Guild to achieve space power? More to the point, how might they achieve FTL travel?

Subluminal Suggestions: Why Do We Even Need FTL?

> Space is big. Really big. You just won't believe how vastly hugely mind-bogglingly big it is. I mean, you may think it's a long way down the road to the chemist, but that's just peanuts to space.
>
> —DOUGLAS ADAMS, *Hitchhiker's Guide to the Galaxy*

There's a reason that space is called *space*. It's generally known that space is vast and largely empty, but that understanding is more often qualitative as opposed to quantitative. Let's explore a few examples to illustrate how truly immense interstellar distances are. On the National Mall in Washington, D.C. is a 1:10,000,000,000 (one to 10 billion) scale model of our Sun and its planets called the Voyage Model of the Solar System. In the Voyage Model, Sol (aka the Sun) is roughly the size of a cantaloupe. At that scale, Earth is the size of a pinhead and is nearly fifty feet away. Jupiter is marble-sized and is over 250 feet from the sun. At that same scale, the nearest star to Sol is the trinary star system Alpha Centauri, which would be represented by a slightly larger cantaloupe than that for Sol (Alpha Centauri A), in mutual orbit with a slightly smaller cantaloupe (Alpha Centauri B), and with a little red golf ball in orbit about the pair (Proxima Centauri). At the scale of the Voyage Model, all three stars would be situated in Golden Gate Park in San Francisco, over 2,500 miles away! That is the *nearest* star system to Sol. The distance from Earth to Caladan (Delta Pavonis 3) is nearly five times that.

Not only are interstellar distances, well, astronomical, but current human technology has not yielded a propulsion system that can cover such distances in a human lifetime. When the Apollo astronauts visited our nearest celestial neighbor, the Moon, in the late 1960s and '70s it took them three days one way. With "modern" technology, we have no reason to believe that the trip would be significantly, if any, shorter (it should be noted that presently the United States does not build a single booster with near the thrust or heavy lift capacity as the Saturn V rocket that took the Apollo astronauts to the Moon). While light can take anywhere from four to twenty minutes to get to Mars—depending upon where Earth and Mars are in their respective orbits—a typical robotic Mars probe takes six months to a year to make the journey.

Looking further out into the Solar System, it took the Cassini spacecraft six years to reach the planet Saturn, but this duration is, admittedly, somewhat misleading. The Titan IIb booster that launched Cassini from Earth did not have the thrust necessary to send the spacecraft on a direct trajectory to the ringed planet, so Cassini performed two flybys of Venus, one of Earth, and one of Jupiter. In each case the spacecraft got a boost of speed from the planet's gravity, thus, in turn, slowing the planet's orbital speed by an infinitesimal amount. This underscores the point that the current state of propulsion technology requires that humans must "get a little help from our friends" to send a six-ton (in Earth's gravity) spacecraft to the outer Solar System. Spacecraft carrying humans would be required to carry food, water, waste storage, life support, radiation shielding, living space, and a whole host of other facilities necessary to keep the crew alive (and relatively emotionally stable) during a long voyage. Such a craft would be far more massive and require far more force to propel it to even modest speeds.

What about interstellar travel, when distances are measured in light years? Contrary to frequent colloquial usage, a light year is not a measure of *time*, but rather a measure of *distance*; the distance light travels about 6 trillion miles in one year. The most distant object ever built by humankind is the Voyager 1 spacecraft. In early 2007, Voyager 1 was slightly more than 102 astronomical units from the Sun, but it took nearly thirty years to travel that distance (one astronomical unit, or AU, is approximately the average distance from Earth to the Sun, about 93 million miles). That distance, 102 AU, is just slightly more than 1/1000 of one light year. Voyager 1 is currently traveling into interstellar space at 3.6 AU per year in the direction of the constellation Ophiuchus, the little-known thirteenth constellation of the Zodiac. At its current velocity, Voyager 1 will pass within 1.64 light years of the small red dwarf star AC+79 3888, currently located in the constellation Ursa Minor, in 40,000 years, give or take a few thousand years. At that speed, were Voyager 1 headed toward Alpha Centauri, it would take between 75,000 and 80,000 years.

What if we are impatient? What if we want to see the universe with our own collective eyes as soon as possible? The realms of both science and science fiction have suggested ways in which humanity may explore the stars, constrained by the ability to travel at fairly pedestrian speeds: sleeper ships and generational ships. A sleeper ship is a vessel in which the inhabitants are placed some form of suspended animation—be it deep hibernation or cryogenic freezing. The goal is to arrive at the destination years later with the same crew who began the journey, but without the burden of having to transport or maintain tens, hundreds, or even thousands of years' worth of food, water, and air. The *Jupiter2* in *Lost in Space*, *Discovery* in *2001: A Space Odyssey*, and even the Golgafrincham ark from *Hitchhiker's Guide to the Galaxy*, are all well-known examples from the realm of science fiction. At present, though, the technology does not yet exist to put living humans in hibernation for such extended periods. Generational ships, by contrast, are described as self-contained ecosystems designed to travel vast interstellar distances where the journey time is significantly longer than a human lifetime. The crew who arrives at the destination will be the descendents of, and hopefully the same species as, the original crew. This is also a common theme in science fiction, having been explored in *Star Trek* (TOS and *Voyager*), *Space: 1999*, *Firefly*, and even Arthur C. Clarke's *Rendezvous with Rama*.

It is interesting to note that although Frank Herbert never explores the concepts of generational or sleeper ships, there exists the framework within the Duniverse—given the geriatric properties of spice—for a hybrid vessel. We can imagine an ark ship where the inhabitants are not put in suspended animation, but rather provided with an ample supply of spice for a long journey. They would travel awake, as they would be the crew of a generational ship. The spice would slow the aging process, and the original crewmembers arrive at the intended destination as with a sleeper ship. Of course there would have to be severe limits placed on procreation during the voyage. This method of traversing the stars of the Duniverse, as with the Guild Heighliners, would rely on the spice of Arrakis (or Rakis since the trip would take so long, the planet's name would mutate in the interim).

Much of the appeal of the Dune novels is the political intrigue, so the use of any kind of ark for travel is impractical from a dramatic standpoint. Driving the last nail home on the use of ark ships in *Dune*, the distances involved between key planets also precludes this. In the opening chapter of *Dune*, we find House Atreides in the process of relocating from from Caladan (Delta Pavonis 3) to Arrakis (Canopus 3). In reality, the distance between these two stars is slightly more than 300 light years (the program I hacked out to determine this was written with a JPL-developed software library called, appropriately, SPICE). Without rapid FTL travel between two planets, all political connection is functionally severed, as Paul Atreides notes in *Dune* when he considers the ramifications of cutting off the supply of spice: "The Guild is crippled. Humans become little isolated clusters on their isolated planets." A sentiment Leto II echoed in *Children of Dune*:

> Without spice, the Empire falls apart. The Guild will not move. Planets will slowly lose their clear memories of each other. They'll turn inward upon themselves. Space will become a boundary when the Guild navigators lose their mastery. We'll cling to our dunetops and be ignorant of that which is above us and below us (168).

Returning to practical terms, if humanity is to explore even the Solar System, let alone the stars, and we don't wish to do it in generational or sleeper ships, we are going to have to pick up the pace a bit. From 1996 to 2002, NASA sponsored a program based at the Glenn Spaceflight Center called the Breakthrough Propulsion Physics Program (B.P.P.P.), a program designed to investigate new areas of physics which could lead to dramatic leaps in our ability to get from here to there quickly. On the B.P.P.P. Web site, it is emphasized that the goal of the program is not to develop technology for traveling faster, but rather to understand the physical principles that can be exploited by future technology to make profound advances in propulsion:

> This research falls within the realm of physics instead of technology, with the distinction being that physics is about uncovering the laws of

nature while technology is about applying that physics to build useful devices. Since existing technology is inadequate for traversing astronomical distances between neighboring stars (even if advanced to the limit of its underlying physics), the only way to circumvent these limits is to discover new propulsion physics. The discovery of new force-production and energy-exchange principles would lead to a whole new class of technologies. This is the motivation of breakthrough propulsion physics research.

While the program found no "Holy Grail"—no breakthrough to take us to the stars soon—and is currently no longer being funded, it did generate sixteen peer-reviewed articles which may shine a light on future directions for research. Interestingly, the B.P.P.P. created a nice Web site that further elucidates all the practical issues and problems with FTL travel, called "Warp Drive, When?"[1]

Given the name of the NASA program, the Breakthrough Propulsion Physics Program, one might ask, "What are some of the boundaries through which we must break to reach the stars in a reasonable time?" The answer would be, of course, "It's all relative." (Note: If the reader has a familiarity with Special and General Relativity, the next two sections may be skipped.)

Special Relativity: Doing the Space-Time Warp

> Nothing is ever as it seems. With appropriate equations I can prove this.
>
> —NORMA CENVA, *Dune: The Machine Crusade*

Although the speed of light in a vacuum appears to be a cosmic speed limit (for now), light—more correctly electromagnetic (EM) radiation—travels at a finite speed. Nothing known to current science can travel faster than the speed of light in a vacuum, at least nothing that has mass or carries with it information. To elucidate what it means to "carry information," since radio waves, a form of EM radiation, can be used to transmit AM/FM radio or television, EM radiation both trans-

---

[1] http://www.nasa.gov/centers/glenn/research/warp/warp.html.

ports energy and carries information. Light travels at 186,000 miles per second, or nearly 300,000 kilometers per second. Light could travel around the circumference of Earth seven times in one second. A beam of light takes about 1.3 seconds to reach the Moon. Sunlight takes nearly 8.5 minutes to reach us.

This last fact underscores a very important point: space and time are inexorably linked. When you peer out into space, you also look back into time. When you view the Sun you do not see it as it is right now, you see it as it was 8.5 minutes ago—the time it took its light to reach your detector which, hopefully, is not your unshielded eye. Our nearest stellar neighbor is the trinary star Alpha Centauri, which is 4.2 light years away. When we see Alpha Centauri, we see it as it was a little more than four years ago. M31, known better as the Andromeda Galaxy, is the most distant object one can see with the unaided eye: it is approximately 2.8 million light years away.

Since it takes 2.8 million years for its light to reach us, when we see M31 we see it as it was when human beings barely resembled anything we would call "human" today. So space and time cannot be thought of as independent quantities: space has three physical dimensions, time adds another. When we consider travel to other stars, we think in terms of traveling through four-dimensional *space-time*. The character Chani alludes to this in *Dune Messiah*: "Denying them their melange would solve nothing. . . . So the Guild's navigators would lose their ability to see into timespace" (28).

Science can be defined as an organized methodology for understanding nature and reality based upon observable phenomena. Science relies on such concepts as objectivity, reproducibility of results, and measurement. On this last point, it was when Albert Einstein published his Special Theory of Relativity in 1905 in an article called "On the Electrodynamics of Moving Bodies" that our collective concept of what can be measured and even the meaning of those measurements was thrown into disarray.

There are two of the postulates at the core of Special Relativity (SR). The first essentially states that there is no such thing as an absolute definition of motion in the universe. Therefore, since everything can be said to be in motion, motion must be specified relative to a reference frame. Perhaps this is best explored using an example. Imagine

FIGURE 1: M31, the Andromeda Galaxy. Image Courtesy NASA/STScI

that we have two observers: Leto and Jessica. For her birthday, Leto has bought Jessica a brand-new space cruiser (it's good to be the Duke) called the *Spirit of Caladan*. Because the ship is not Guild approved, the *Spirit of Caladan* can travel nearly the speed of light, but not beyond, and is designed merely for cruising around the Canopus system. Excited about her new toy, Jessica immediately insists on taking it for a spin. Tired from a long day of foiling Harkonnen plots, Leto is content to sit by and watch. The Duke retires to his observation satellite in space above Caladan while Jessica, eager to show her appreciation, has the crew the *Spirit of Caladan* pass Leto at a uniform speed. The Duke, stationary as far as he can tell, smiles to himself as he appreciates the beautiful ship flying past, sheer poetry in motion. Meanwhile Jessica, seated on her new ship, waves as the Duke's observation station passes and recedes into the distance. From the framework of SR, both observers can be considered equivalent. In other words, while Duke Leto felt he was the stationary observer and the *Spirit of Caladan* was the moving object, Jessica would be equally correct to say that she was at rest and it was Leto who flew past, albeit in the opposite direction. Motion is gauged relative to a given *reference frame*.

The second postulate of Special Relativity is that the speed of light in a vacuum is invariant, a universal constant, and independent of the motion of either the source of the light or the receiver. Initially this postulate is rather simple to grasp, but its implications are quite counterintuitive. Imagine a railroad flatcar is traveling at sixty miles per hour. Further imagine the unlikely scenario that two people, with one on either end of the flatcar, are playing a game of catch: they are tossing a ball back and forth with a speed of five miles per hour one way then the other relative to each player. The players would only ever measure the speed of the ball to be five miles an hour. Now imagine an observer who is stopped and waiting at a railroad crossing, watching them play catch as the flatcar passes by. If that observer were to measure the speed of the ball, it would vary between sixty-five miles an hour (when the ball was thrown from back to front) and fifty-five miles an hour (when the ball was thrown front to back). That is rather intuitive.

Now imagine that the two players on the flatcar were playing laser tag, shooting beams of light back and forth. The players would measure the speed of the laser light to be 670,616,629 miles per hour one way, then the other. The observer waiting at the railroad crossing would also measure the same speed for the light beam as the two players. How could that be? It turns out that if the players measured the distance between them, they would measure one value. If the observer at the crossing was able to measure that same distance, it would be slightly smaller, even though the flatcar was moving. Further, if the observer were particularly observant, he would notice that the wristwatches of the two players were running...ever...so...slightly slower than his. So the one constant throughout this example was the speed of the light beams and the measurements of distances and times varied—dependent upon who was moving relative to whom—to ensure that the speed of light was the same for all observers. The discrepancies between the measured times, distances of the players, and the stationary observer would be even more pronounced if the flatcar was moving a significant fraction of the speed of light.

Now relativistic changes to time, space, and mass from near-luminal and FTL travel have been almost universally ignored in science fiction—at least those implied by SR. Let us, however, briefly examine how some of these effects would impact the development of any future FTL drive.

Relativistic or Apparent Mass

There are three commonly used definitions of mass. The colloquial definition that was taught to most high school and college students is that an object's mass is a measure of how much matter the object possesses, a count, if you will, of the sum total of all the electrons, neutrons, and protons that compose the object. In short, mass is a measure of how much "stuff" an object contains.

An object's *mass* is not to be confused with its *weight*. If you multiply an object's mass by the pull of gravity it experiences, you calculate its weight. Suppose you travel to Earth's moon. The strength of the moon's gravity is roughly 1/6 of Earth's, hence your weight on the moon is 1/6 that of your weight on Earth. If you were in intergalactic space, where the pull of gravity is very small, you would be nearly weightless, but your mass would still be constant in all cases. If, however, you are back home on Earth and go on a strict beer and meat lover's pizza diet, your mass is almost certain to increase.

Students taking physics at the college level are exposed to a different definition of mass. An object's mass is its resistance to a change in motion when a force is applied. An automobile has more mass than a shopping cart; therefore, anybody who has visited a grocery store and used a cart and has also pushed a broken-down car knows that it takes significantly more force to push the automobile than to push the shopping cart. More germane to the topic at hand, thrust—as in the thrust from a jet or rocket engine—is a measure of force. The more massive a spacecraft is, the more thrust it takes to propel it.

The definition of mass used by many disciplines within physics is again different: an object's mass is defined solely as its energy equivalent. The question is begged, "What is energy?" Energy can be defined as the ability to move an object that has mass. When physicists speak of energy, we find that there are two basic forms: potential energy and kinetic energy. Any moving object has kinetic energy; an object at rest has zero kinetic energy. We can view potential energy as the "potential to have kinetic energy." If you hold a rock at arm's length, you know that if you let go, gravity will cause it to accelerate to the ground and potential energy will be converted to kinetic. On the other hand, if you throw the rock straight up, the rock will have increasing potential the higher it goes, but gravity will cause it to slow to a stop, then fall back to Earth. Clearly we see how kinetic energy can be converted to potential energy and back. In all these cases, the total amount of energy, the sum of the rock's kinetic and potential energies, is a constant. Physicists say that it is *conserved*.

Just as kinetic energy and potential energy can be converted, one into the other, the most famous equation in all of physics tell us that mass can also be converted into energy:

$$E = mc^2$$

This equation says that if you could convert the entire mass of an object (m) entirely into energy, the energy output would be equivalent to the product of the mass times c, which is the speed of light squared. The speed of light is a very large number, that number squared is a huge number, therefore even a small amount of mass yields a huge amount of energy if it's converted. By way of example, if you were able to convert fifty-five grams of mass (slightly less than two ounces) completely into energy, it would have the explosive yield of a 1 MT nuclear explosion—where 1 MT means one megaton, or the explosive equivalent of 1 million tons of TNT. Every second, our sun converts 4 million tons of matter directly into energy. That energy comes to us in the form of sunlight.

As an object is accelerated to increasingly higher speeds, its mass, or its resistance to an additional increase in speed, *appears* to increase according to the equation:

$$m_{relativistic} = \frac{m_{rest}}{\sqrt{1 - \frac{v^2}{c^2}}}$$

Where m_{rest} is the object's mass when it is stationary, c is the speed of light, and v is the object's speed. The object acts as if its mass ($m_{relativistic}$ in the equation above) increases with increasing speed. Physicists no longer use the concept of relativistic mass, preferring instead to concentrate on the sum total of an object's energy. Nevertheless, it is a useful construct when explaining the difficulties of accelerating a body to near-luminal speeds since it demonstrates how an object's reaction to a force like thrust decreases as its speed increases. It also clearly demonstrates impossibility of attaining light speed.

If we are to plot the above equation, it would look this:

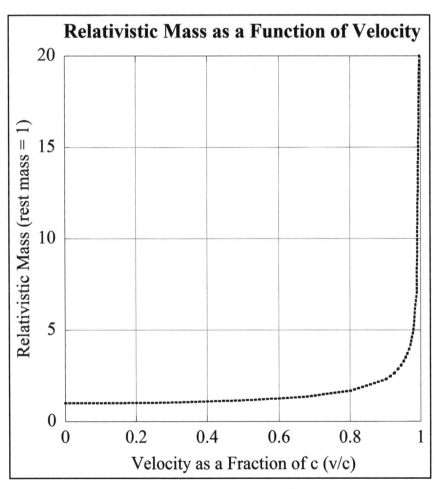

FIGURE 2: "RELATIVISTIC MASS"

As the velocity approaches the speed of light, the term v^2/c^2 approaches the value 1. The term $1-v^2/c^2$ approaches 0. The square root of zero is zero, and the result of any number divided by a number approaching zero approaches infinity.

Let's return to our example of the *Spirit of Caladan*. Jessica is again aboard the ship and Leto watches from his observation platform in space. The *Spirit* moves off to a safe distance, turns, then fires its engines, accelerating past Leto. If Leto was somehow able to measure the *Spirit*'s resistance to any change in motion, it would increase in the manner shown in Figure 2. As a force (like thrust) is applied to an object (like the *Spirit of Caladan*) to increase its speed, an even greater force is needed to further increase its speed by the same amount. The speed of light is not attainable.

Two things should be noted. First, recall that our second observer, Jessica, is aboard the *Spirit of Caladan*. For her the spacecraft is functionally at rest, therefore as the craft speeds rapidly past Leto, its motion doesn't change relative to her. In other words, since the *Spirit of Caladan* appears to be at rest relative to Jessica, it takes her no more or less effort to move about the spacecraft as it did before it was in motion.

Second, although the speed of light is unattainable, there is nothing within the realm of SR that says an object cannot move *faster* than the speed of light. The speed of light is a barrier, but if there is a way to travel FTL without actually traveling at the speed of light at some point—a way to "tunnel" through the light barrier—perhaps superluminal travel is possible. Science fiction is rife with references to *tachyons*, particles that only ever travel FTL. The unfortunate thing about tachyons is, as we shall see from other equations of SR, their annoying habit of traveling backward in time.

Time Dilation

Time. We always have too little, or too much—never just enough.

—Norma Cenva, *Dune: The Machine Crusade*

Another outcome of SR is that a clock in motion moves more slowly than a stationary clock. This is called relativistic time dilation and, similar to the equation for relativistic mass, the equation for time dilation is:

$$\Delta t_{moving} = \frac{\Delta t_{rest}}{\sqrt{1 - \frac{v^2}{c^2}}}$$

Let's return to Leto and Jessica. As Jessica flew past her Duke at half the speed of light in the *Spirit of Caladan*, a sharp-eyed Leto may have noticed that for every ten seconds that elapsed on his watch, a little more than eleven and a half elapsed on hers. On another flyby, this at 99 percent the speed of light, for every ten seconds that pass on Leto's watch, over a minute and ten seconds pass on Jessica's. One counterintuitive aspect of time dilation comes when we recall that all references frames are equivalent. From her standpoint it was Leto's watch that was slower each time Jessica flew past the Duke.

The effect of relativistic time dilation was actually confirmed experimentally by researchers Rossi and Hall in 1941. *Muons* are subatomic particles that are created when cosmic radiation interacts with Earth's upper atmosphere. Although muons travel very rapidly, they also decay into smaller subatomic particles in slightly over two-millionths of a second on average. Given the traveling speed and brief lifetimes of muons, one might expect that a muon might travel a few hundred meters at most before decaying. In fact, numerous muons reach the surface of Earth. This is because muons travel at nearly the speed of light, their lifetimes are extended due to time dilation, and a significant fraction of the muons actually reach Earth.

Another counterintuitive aspect of time dilation is that an object with mass, traveling FTL, would travel *backward* in time.

Lorentz-Fitzgerald Contraction

Having moved to Arrakis years after Leto's and Jessica's first experiences with relativity with the *Spirit of Caladan*, Leto decides that Jessica is due for an upgrade. He buys her a new vessel, the *Arrakeen Queen*. Learning that they had overlooked one particularly interesting effect of SR, they again arrange for a series of flybys, for word has it that the faster Jessica flies past, the shorter her ship will become. Both are dubious, but they decide to try the experiment anyway. Before the experiment begins, both Leto and Jessica measure the length of the *Queen* and agree that at rest it is, coincidentally, one *shai-hulud* in length (because the unit of length and distance on Arrakis may very well be measured relative to the average length of a sandworm).

The *Arrakeen Queen* flies past Leto several times, each time at increasingly greater speed. At low speed, both Leto and Jessica would still agree that the Heighliner is one shai-hulud in length:

For subsequent flybys, though, an odd thing starts to happen. Leto notices that for each successive flyby, the *Arrakeen Queen* seems to be getting shorter. In fact, if the length of the *Queen* at rest is L_{rest}, then the length that Leto would measure follows the equation:

$$L_{moving} = L_{rest}\sqrt{1 - \frac{v^2}{c^2}}$$

The second pass is at half the speed of light, and Leto notices that it is now compressed to 87 percent of one shai-hulud:

At 0.9c the effect is now quite pronounced:

Finally, at 0.99c, which is top speed for the *Queen*, Leto would see:

This effect is called *Lorentz-Fitzgerald Contraction*, or often just *Lorentz Contraction*. We see this plotted graphically in Figure 3.

When Leto and Jessica are reunited, Leto tells her the exciting news—that her ship was, indeed, "squashed" as it passed. A confused Jessica, naturally, has no idea what he's talking about, since it never changed size for her. Since she was at rest relative to the ship, it remained the same size, though she admitted that her Duke looked a bit funnier each successive pass. Recall that from Jessica's reference, she was stationary relative to the ship while it was Leto who was moving past at increasingly higher speeds.

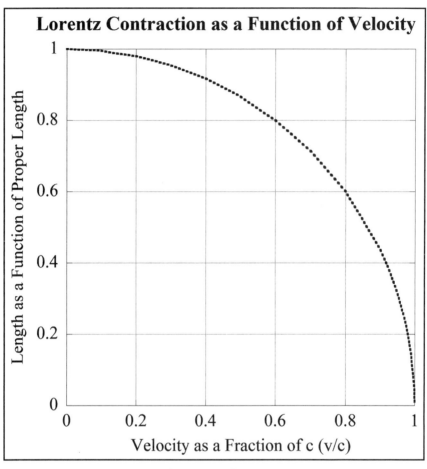

FIGURE 3: LORENTZ CONTRACTION

It is also important to note the importance of this result on potential FTL travel within the equation for Lorentz Contraction. Lorentz Contraction implies that physical distances are not absolutes—that the meaning of physical distances can be different depending on the state of the observer and if there are instances when space can be compressed.

All the relativistic effects we have explored, in addition to complicating the possibility of FTL travel, have interesting and unusual side effects. Imagine that we have twins living on Arrakis, Leto II and Ghanima. The twins hear a rumor that Earth, previously a smoking cinder after the Butlerian Jihad, is now marginally habitable. Rumor also has it that the Guild are breeding sand trout on Earth, trying to create another Arrakis. This would hit at the very heart of the power of House Atreides, so the twins decide that Leto II will visit Earth to learn the truth of matters while Ghanima remains on Arrakis. Since the Guild can't learn of this visit, Leto II brings their grandmother's cruiser, the *Arrakeen Queen*, out of the mothballs for the journey. Of course, given that Earth is 310 light years from Arrakis, this might take some time. Thank the little makers for the spice!

Imagine that the *Queen* accelerates rapidly to its top speed, 0.99c, and coasts most of the way to Earth. During the long and often boring trip we can envision the crewmen—like Earth sailors of days gone— passing the long hours drinking spice beer, singing space shanties, and sharing bawdy limericks:

There was a Harkonnen called Nisk,
Whose stroke was exceedingly brisk.
So quick was his action,
That Lorentz Contraction,
Reduced his shaft to a disk.

Assume that the acceleration and deceleration times for the trip are negligible. Further assume that the stay at Earth is very short (Leto II learns that Earth is still, in fact, a smoking hole in space). Since Leto II is traveling near the speed of light, the distance between Earth and Arrakis is significantly shortented—only 43.7 light years. The round trip time for Leto II ends up being a "mere" eighty-eight years, three months. Thank the little makers for the spice! While Leto II is gone, 626 years and three months elapse on Arrakis, however. No amount of spice in the Duniverse could have prolonged his twin's life for that long, and Leto II returns to meet Ghanima's great[17]-granddaughter. If it wasn't before, it now becomes crystal clear why relativistic effects are largely ignored in science fiction, and why the concept of FTL travel is a necessity for most stories.

Causality

However carefully you plan for the future, someone else's actions will inevitably modify the way your plans turn out.

—JAMES BURKE, *Connections*

While certain physical phenomena can, in fact, travel FTL, nothing which has mass or carries information can. This leads to interesting definitions of *causality* when we consider relativity. If one physical event (the cause) directly leads to another (the effect), then we can say that the first event caused the second. Imagine the two events are widely separated, however. If the first event happens at point A and the second at point B. If the distance between points A and B is so great that, in the intervening time between the two events, a beam of light has insufficient time to make the journey between them, we say that the event at A could not have caused the event at B. If a beam of light can connect the two events, then the event at A *may* have caused the event at B.

In the Duniverse we learn that "melange precipitated the 'navigation trance' by which a translight pathway could be 'seen' before it was traveled" (*Children of the Dune* 84). The concept of prescience, and the ability to see the "safe" path across many light years, certainly violates the relativistic law of causality. In order to see the safe path, the Guild Steersmen would have to glean details, more or less instantaneously, across the space of an entire journey.

General Relativity: Let's Do the Space-Time Warp Again

Over the span of 1915 through 1916, Albert Einstein published two papers expounding the General Theory of Relativity (GR). Prior to GR, Special Relativity seemed to contradict Newton's Universal Law of Gravitation. GR reconciled the two theories. A fundamental postulate of GR is that the force of gravity is actually not really a force at all, but rather a bend or warp in the fabric of the space-time continuum. Any object that has mass curves space-time. More mass means more of a curve.

A good way to understand this is to imagine a sheet pulled taut over a bed. Now throw a ball bearing onto the sheet. There will be an indentation where the ball bearing rests. That represents the ball bearing's gravity. Toss a softball onto the bed and it will make an even larger indentation. More mass means more gravity. Toss a bowling ball onto the sheet to get a huge indentation and you see that a lot of mass means a lot of gravity. If you then rolled the ball bearing past the bowling ball, the indentation cause by the bowling ball would alter the trajectory of the ball bearing if the two passed close enough. This is equivalent to a passing object being deflected by the gravity of a star. If the ball bearing is not rolling fast enough, it may even fall into the indentation created by the bowling ball, thus having been captured.

Prior to GR, Newton's theory of gravity required that both objects have mass in order for there to be a mutual gravitational attraction. An implication of GR, though, is that if large mass like a star curves spacetime, then anything passing by the star—including light—would have its trajectory altered.

This bending of light, predicted by GR, was first observed experimentally in 1919 by Sir Arthur Stanley Eddington. Traveling to an island off the west coast of Africa, Eddington photographed a total solar eclipse as well as stars in the line of sight of the Sun during the eclipse. The apparent positions of the stars near the Sun were shifted by the amount predicted by GR. When spacecrafts need to communicate with their controllers on Earth—to "phone home" as it were—they broadcast electromagnetic (EM) radiation in the radio portion of the EM spectrum (visible light is a form of EM radiation that we can see; radio is simply a different "flavor" of EM radiation that we can't see) to Earth. if the spacecraft is on the far side of the Sun relative to Earth, the Sun's gravity bends the radio transmissions, and must be taken into account (Figure 4).

As telescopes like Hubble, Spitzer, Keck, and VLT are able to peer increasingly deeper into the Universe, scientists increasingly see effects directly attributable to GR. One of these effects is called *gravitational lensing*. Imagine that we are observing a very distant object; one at the very edge of the known universe, like a quasar. Imagine further that there is a very massive object, like a galaxy or even a black hole, between the observer and the quasar. We would expect the massive

FIGURE 4: Image Courtesy NASA/JPL/Caltech

object to bend the distant light of the quasar, but sometimes that light takes multiple paths, as is shown in Figure 5.

The gravity of the massive object can act as a lens. If the source of the light, the massive lensing object, and the observer all lie on a straight line, this may not only cause the source object to appear in a shifted position, but the light of the source may take multiple paths, creating multiple images, as seen in Figure 6.

Gravitational Lensing Splits Quasar Light into Five Images

Distant quasar with host galaxy

Light emitted from quasar bends around intervening galaxy cluster, producing lensed images*

*The red crescents represent lensing arcs — Smeared images of background galaxies.

FIGURE 5: Graphic Courtesy NASA/STScI

If GR tells us that space can be curved, bent, or warped, then perhaps it can be compressed or folded. With what we now know about SR, GR, and the fabric of space-time, we now understand the basis for how FTL has been represented in science fiction over the years.

FTL As Implemented in SF

FTL travel has been depicted in science fiction many ways, but generally falls into four basic categories, with varying degrees of scientific validity:

Einstein, Schmeinstein, We're Sticking with Newton

Some stories have taken the non-physical approach that SR can be ignored, and Isaac Newton's laws of motion work just fine up to the speed

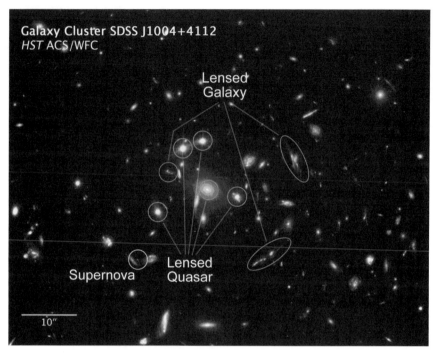

Galaxy Cluster SDSS J1004+4112
HST ACS/WFC

Lensed
Galaxy

Supernova   Lensed
Quasar

10"

FIGURE 6: Hubble Image Courtesy NASA/STScI

of light, even beyond. One excellent example of this is in the classic *Battlestar Galactica* episode "The Living Legend" (Episode #11). Regarding a major battle with the Cylons, Commander Adama tells Commander Cain of the Battlestar *Pegasus*, "If you don't mind me using up half our fuel, I'll even see if I can't bring her (*Galactica*) up to light speed." With an overwhelming body of evidence supporting the effects described by Special Relativity, we know that it would take vastly more than half of *Galactica's* fuel to accelerate to even a large fraction of light speed. While convenient from a storytelling standpoint, many science fiction fans have enough of an idea that odd things happen when traveling at near-luminal speeds that this device is used increasingly less often. In fact, acting in the capacity as science advisor for four seasons of *Battlestar Galactica* v. 2.0, I would not let a line like that slip past without a deluge of complaints. Such a departure from anything that even remotely resembles reality is easily recognized by the audience, and a work of fiction employing this device is easily dismissed. That is, when well-known physical laws are completely ignored in a work of so-called *science* fiction, the attention of the technically literate among the audience—and the credibility of the work—is damaged, perhaps entirely.

Drive the Instellar Autobahn, Where There Is No Speed Limit

A common device used in science fiction is the creation of an alternate dimension where space is "denser," the speed of light is not the universal speed limit, or there is no speed limit at all. A spacecraft enters this dimension through its own or external means (by the way of a jump gate versus a jump point in *Babylon 5* terminology), it travels rapidly to the destination, then reemerges into "normal" space. Such a dimension has been given names like hyperspace (*Star Wars*, *Babylon 5*, and numerous other works), subspace (*Star Trek*, though the plot device is used for communications only), or slipspace (*Andromeda*, *Star Trek: Voyager*, and the Halo series). When hyperspatial travel initially appeared in science fiction in the 1930s, there was no corresponding science to explain it, even until recently.

Perhaps recent avenues of thinking in the areas of cosmology, string theory, and M-theory, a concept called *brane cosmology*, can provide a post-hoc explanation for hyperspace.

The fundamental premise of brane (short for membrane) cosmology is that our four-dimensional space-time may be one of many (normally) disconnected universes. These universes coexist simultaneously within a higher dimensional space often called the *bulk* (Figure 7). Together all the parallel universes along with the bulk form what has been called the multiverse. The idea that there may be multiple universes coexisting simultaneously has been used frequently in science fiction, though the concept of multiple universes and multiple realities goes back much further and can even be found in ancient Hindu writings. In fact, the concept that there are disconnected parallel universes, immersed in a greater expanse, was explored in *Doctor Who* (tenth Doctor). The interstitial space between galaxies—again, what cosmologists call "the bulk" was, in that case, referred to as simply "The Void," though other cultures called it "The Howling" or "Hell."

Not only is there is no guarantee that the physical laws that prevail in our universe are the same in others, there is no guarantee that our physical laws would rule in the bulk. Perhaps, then, hyperspace can be

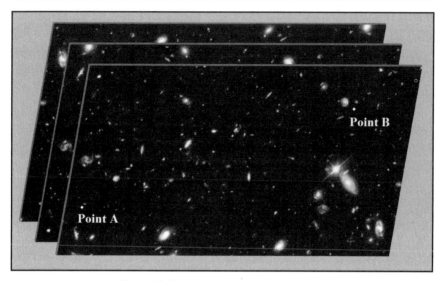

FIGURE 7: BRANES WITHIN THE MULTIVERSE

considered to be the interstitial space between branes, the void between universes. Perhaps there the speed of light has a different meaning...or no meaning. With these caveats, perhaps the science fiction concept of hyperspatial travel really implies that a spacecraft has the capability of exiting its normal universe at Point A, traveling through the bulk at, perhaps, much greater speeds, and re-entering its normal universe at a pre-calculated location (Point B). Great big question marks in physics like this have always been welcome signposts for science fiction writers.

Find a Shortcut

Like hyperspace, another concept that has been used frequently in science fiction to explain FTL travel is that of the wormhole (e.g., *Stargate SG-1*, *Star Trek: Deep Space Nine*), a shortcut through space-time. Unlike the concept of hyperspace, these theoretical shortcuts had an earlier genesis, and arise from mathematical solutions that describe Einstein's equations of General Relativity. The term *wormhole* was first used in 1957 by the physicist John Wheeler to describe shortcuts through space-time. Imagine a worm wanted to get from point A on the outside of an apple, to point B on the far side. The worm could either travel around the outside of the apple, or chew a shorter path through the apple.

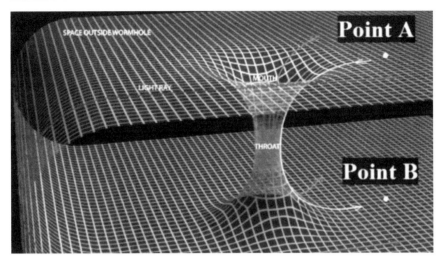

FIGURE 8: GEOMETERY OF A WORMHOLE
FIGURE ADAPTED FROM HTTP://COMMONS.WIKIMEDIA.ORG/WIKI/IMAGE:WORM3.JPG

Figure 8 shows how this might be applied to FTL travel, again between our points A and B. Figure 7 displayed two-dimensional sheets to represent the three special dimensions of our universe, our "brane." GR says that space can be curved or warped. So what if we are able to fold that sheet—the fabric of our space-time—and then find a way to "tunnel" between point A and B, akin to the way our worm chewed through its apple? The distance between the points would be drastically reduced, and we could have functionally traveled FTL (without, perhaps, ever actually traveling FTL within the wormhole itself), simply by virtue of a shortened path.

The first wormholes that were discovered mathematically were microscopic and short-lived in nature. Even then, they were very unstable, and collapsed when matter entered them. Several such solutions have been discovered over the years, but it was in 1988 that researchers discovered wormholes that could be both stable enough and large enough for a spacecraft to traverse.

Distort the Space-Time Fabric

As we have already established, both theory and observations of General Relativity make it clear that space-time can be curved, warped, perhaps even folded or compressed. One FTL plot device in science fiction, then, has been the depiction of a drive that can compress, or warp, space ahead of the craft (e.g. *Star Trek, Battlestar Galactica* v. 2.0), thus functionally creating a shorter trip, or even a "jump." Similar to wormhold travel, the spacecraft may not ever surpass the speed of light in its local environment, but by significantly shortening the distance between the beginning and end of the journey, it functionally travels faster than light.

Hybrid Technologies

In *Star Wars: A New Hope,* Han Solo claims that his ship, the *Millennium Falcon,* is the fastest around when he brags, "Is she fast? It's the ship that made the Kessel run in less than twelve parsecs." This statement can be interpreted in one of two ways. Since a *parsec* is a measure of distance (3.26 light years), yet in context Han's usage implies a measure of time, this could mean that his statement was mere bluster. Alternately, this can be seen to imply that the *Falcon* uses a marriage of two technologies. We know that the *Falcon* ostensibly travels through hyperspace to achieve FTL. Does the ship additionally *fold* hyperspace? In the Star Wars universe, is the machismo of one's hyperdrive measured by whose drive can bring distant points the closest? I lean toward the "bluster" argument, but this apparent conflict is still debated among Star Wars fans (restated, I have actually witnessed fans having this debate more than once at science fiction conventions, and the "*Star Wars*" episode of the television series *Family Guy* also alluded to this paradox). Other works have hinted that FTL in their realm is a convolution of the plot devices listed above.

FTL in Dune

In the first trilogy of Dune novels, there is very little information about the actual science behind the drives that propel the Heighliners at translight speeds. In fact, Frank Herbert did little in the way of prophesying the technology that would exist at the time of the imperium. Perhaps that was on purpose. If you make no predictions, nothing ever becomes dated, thus ensuring your work is a classic for the ages.

The reader learns only that FTL in *Dune,* in fact much of the technology, is based upon the ill-defined Holtzmann (or Holtzman) Effect. This effect is the basis for personal shields, suspensors, and glow globes. In the Terminology of the Imperium we have the following definitions:

HOLTZMAN EFFECT: the negative repelling effect of a shield generator.
SUSPENSOR: secondary (low-drain) phase of a Holtzman field generator. It nullifies gravity within certain limits prescribed by relative mass
and energy consumption.

We know that Guild Navigators require the spice melange to pilot the
ships through folded space, but little more. It is only with the later books
that a clearer picture emerges. It appears that FTL technology in the Duniverse is a hybrid of the fourth option above: the compression or warping
of space-time along with another technology: mass neutralization.

There is the implication in the Dune novels that Heighliners generate a Holtzmann field, which neutralizes the ship's mass to some
degree. Recall how rapidly an object's apparent, or relativistic, mass
increases as a function of speed. Also recall how, at high speeds, a
dramatic increase of thrust is required for a comparatively small increase in speed. If the mass of the spacecraft could be "nullified" by
a Holtzmann field, then this is one relativistic effect that can be minimized. The scientific basis behind mass neutralization is discussed further in Ges Seger's essay "Suspensor of Disbelief" (page 207) and will
not be expounded upon further here.

It wasn't until the fourth Dune novel that we began to learn more about
the history of FTL in the Duniverse. Said by Leto Atreides II, in *God Emperor of Dune*, "You think a man designed the first Guild ship? Your history books told you it was Aurelius Venport? They lied. It was his mistress,
Norma. She gave him the design, along with five children" (183).

In the (non-canon) Legends of Dune books we found that Norma Cenva learned the secret of folding, or compressing, space-time.
This allowed large vessels—later called Heighliners—near-instan-
taneous travel through foldspace and across light years of normal
space. Norma and her lover, Aurelius Venport, created the first
shipyard on the planet Kohlar to build ships that would navigate
foldspace. At the onset, it became obvious that navigation through
foldspace was dangerous, and not all of the ships that used their FTL
drives arrived at their destination. It was again Norma Cenva who
discovered that the prescient abilities given to one who has ingested the spice melange allowed safe navigation through foldspace and
across the Duniverse.

In his novel and television series *Connections*, author James Burke explores the history of technical innovations that have led to dramatic unintended consequences. Irrespective of the technical implementation of space-folding in the Duniverse, one should be impressed with the *Connections*-style creativity of Frank Herbert in creating a universe where excrement from larval sandworms leads ultimately to titanic Heighliners leaping light years across the galaxy.

KEVIN R. GRAZIER, PH.D., is a planetary scientist at NASA's Jet Propulsion Laboratory (JPL) in Pasadena, CA, where he holds the dual titles of Investigation Scientist and Science Planning Engineer for the Cassini/Huygens Mission to Saturn and Titan. There he has won numerous JPL- and NASA-wide awards for technical accomplishment. Dr. Grazier holds undergraduate degrees in computer science and geology from Purdue University, and another in physics from Oakland University. He holds an MS degree in physics from, again, Purdue, and he did his doctoral work at UCLA. His Ph.D. research involved long-term, large-scale computer simulations of Solar System evolution, dynamics, and chaos—research which he continues to this day. Kevin is also currently the Science Advisor for the PBS animated series *The Zula Patrol*, and for the Sci-Fi Channel series, *Eureka*, as well as the Peabody-Award-winning *Battlestar Galactica*.

Commited to astronomical education, Dr. Grazier teaches classes in stellar astronomy, planetary science, cosmology, and the search for extraterrestrial life at UCLA, Cal State LA, and Santa Monica College. He has served on several NASA educational product review panels, and is also a planetarium lecturer at LA's landmark Griffith Observatory.

He lives in Sylmar, CA—and occasionally Mesa, AZ—with a flock of cockatiels and a precocious parrot.

References

Douglas Adams. *Hitchhiker's Guide to the Galaxy*. London: Pan Books, 1979.

Einstein, Albert. "On the Electrodynamics of Moving Bodies." *Annalen der Physik*. 17:891–921, 1905.

The Halopedia: http://halo.wikia.com/

Brian Herbert and Kevin J. Anderson, *Dune: The Machine Crusade*, CITY: Tor Books, 2003.

Frank Herbert. *Dune*, CITY: Chilton Books, 1965.

Frank Herbert. *Dune Messiah*, CITY: Galaxy Publishing Corporation, 1968.

Frank Herbert. *Children of Dune*, CITY: Putnam Publishing Group, 1976.

SUSPENSOR OF DISBELIEF

Ges Seger
with Kevin R. Grazier, Ph.D.

From flying carpets to magical acts, to Star Trek's *anti-grav units, we've been intrigued by levitation for centuries. In* Dune, *levitation is done using devices called suspensors. Ges Seger examines methods by which levitation may be technologically possible—as well as some surprising potential side-effects.*

NOT BEING CONTENT with merely losing myself in fictional worlds, for as long as I can remember I have tried to puzzle through just how one might get any type of technology in science fiction to work. Single crystal diamond spacecraft hulls? Child's play. Space elevators? Piece of cake! Lightsabers? A bit tougher. Anti-gravity? Hmm.... Now that may take some doing. Anti-gravity technology has been a staple of science fiction for nearly as long as the genre has been popular. It's not often I get to exercise my background in physics these days when computer hacking is paying the bills, so you can imagine my excitement when given the opportunity to speculate on how *Dune's* suspensor technology might work.

From Frank Herbert's own *Dune: Terminology of the Imperium,* the Holzmann Suspensor is defined as a device utilizing "the secondary (low-drain) phase of a Holtzman field generator to nullify gravity within certain limits prescribed by relative mass and energy consumption" *Dune,* the *Terminology of the Imperium* 513. Also from the same source, the primary phases of a Holtzman field generator can be used as a shield against both guns and directed energy weaponry, or to travel between the stars. Since the key phrase in Herbert's definition of a suspensor would be "nullify gravity," let's focus upon what that may entail.

The concept of mass is described in much more detail in the chapter on faster-than-light travel (page 177) but, succinctly stated, it can either be viewed as the amount of matter an object contains (a count, if you will, of the total number of electrons, neutrons, and protons in an object), or it can be viewed as an object's tendency to resist any change in motion—from experience we know that the more mass an object possesses, the more difficult it is to move. Anybody who has pushed both a stroller and a stalled automobile can attest to this. This latter definition is often referred to as an object's *inertial mass*.

An object's mass, however, should not be confused with its *weight*. The weight of any object is a product of both its mass and the gravitational field acting upon it. So your weight is determined by how much mass you have, and of Earth's gravitational field. The strength of Earth's gravitational field is, in turn, dictated by how much mass it has. If you were resting upon, say, the moon—which has much less mass, hence a gravitational field only one-sixth as strong as that of Earth's— your weight would be only one-sixth of what it is presently (unless you are not on Earth and this book has succeeded beyond anybody's wildest dreams).

In the previous example, we have alluded to an additional two ways in which we can view the concept of mass. An object's *passive gravitational mass* is a measure of the strength of its interaction with a gravitational field. Within the same field, an object with a smaller passive gravitational mass experiences a smaller force, hence has less weight, than would an object with a larger passive gravitational mass. In the weight example above, your mass would be considered passive gravitational mass. An object's *active gravitational mass* is a measure of the strength of the gravitational field it creates. Again, in the example above, the strength of Earth's gravitational field would be directly proportion to its active gravitational mass.

Although it can be convenient to view mass in different ways—inertial mass, passive gravitational mass, active gravitational mass—at some level these terms are all generally assumed to be equivalent. Sir Isaac Newton, who first attempted to postulate the laws that govern gravity, made it clear that any two objects in the Universe that possess mass attract each other via the force of gravity. Even the book in your hands exerts a small gravitational influence. When Earth and Mars are at their closest, Mars has only fifty times the gravitational attraction upon you as this book. At their farthest, that drops to a little under five times the attraction. Gravitational strength also decreases rapidly as the distance between two objects increases.

Albert Einstein refined the concept of gravity in the early twentieth century. Einstein put forth, through his General Theory of Relativity, the idea that two objects with mass do not so much interact through a force called gravity that propagates through space, but rather any object with mass bends and warps space around it—the larger the mass, the greater the warp. So instead of Earth pulling you downward with an invisible force called gravity, essentially Earth's mass creates a three-dimensional hole in space-time, and you're stuck at the bottom. Have a difficult time visualizing that? You're not alone. Einstein's General Relativity assumed that an object's inertial mass and its gravitational mass were equivalent (it should be pointed out that, in some scientists' minds, the jury is still out on this).

So it could be argued, then, that to "nullify gravity," one could either invent a way of dramatically decreasing an object's (passive gravitational) mass, or create a way of generating a force that opposes gravity (a result of the local active gravitational mass). What if a Holtzman field generator nullified gravity by somehow screening the passive gravitational mass of an object so that it is not attracted to nearby object—such as planets? How might that be implemented? The method which springs to mind involves two concepts: the equivalence of gravitational and inertial mass as postulated in Einstein's theory of General Relativity, and the definition of inertial mass relative to the rest of the Universe as postulated in Mach's Principle.

To understand Mach's Principle it is first useful to understand Einstein's Special Theory of Relativity. Although this is another topic that is elaborated upon in some detail in the chapter on faster-than-light travel, succinctly put, Special Relativity postulates that absolute motion in the Universe can not be detected. In other words, there is no universal state of rest—an object's motion is always measured with respect to another object. If two Guild Heighliners are flying past each other, are they both moving, or is one fixed and the other one moving? The answer, in either case, is "yes," but it depends upon your reference frame. There is an old joke that a student approached

Albert Einstein on a train, and said, "Dr. Einstein, does New York stop at this train?"

Mach's principle, simply stated, says that the inertia of a body (and thus its inertial mass) arises from its relationship to the totality of all other bodies in the Universe. If, according to Special Relativity, all motion is relative, how do you measure an object's inertia or its momentum? Inertia, then, results as some form of interaction between bodies, or between a body and the Universe. Einstein stated, "inertia... originates in a kind of interaction between bodies" (Einstein letter). Hypothetically, if you had some way of isolating an object from interacting with the rest of the Universe (never mind how for right now, this is a thought experiment), then perhaps you can lower its inertial mass. If gravitational and inertial mass are indeed equivalent, you've just altered its gravitational mass as well.

Since the Holtzman field does such a good job screening most other forms of energy, perhaps its low-drain phase somehow screens the mass of the object within it from the rest of the Universe. Somehow lower the mass of the object being attracted, and you lower the force of gravity upon that object. Lower it enough, and it becomes trivial to hover above a planetary surface. A pretty idea. Now let's look at the scientific consequences. For the sake of this argument, let's assume that my son Luke has built a Holtzman mass neutralizer for his school science fair project. Let's also assume it's sitting on the kitchen table in my home just outside Dayton, OH. I tell him to switch it on. He does. Luke's project assumes a mass so negligible that it might as well be zero. The celebration is very short-lived. Why? What happened?

If you answered, "Why it would float above the kitchen table," you're right—for a tiny fraction of a second. In reality, the fun has just begun. As have the lawsuits. There are the small details of these little physical laws called *conservation of momentum* and *conservation of energy* that have heretofore gone unmentioned, but which now rear their ugly heads. Briefly stated, the total momentum of an object having no interactions with an external agent (such as a force acting upon it) is a constant value. Similarly, the total energy of an object (rest, potential, and kinetic) in the absence of some sort of external agent must also be a constant value. Luke's hypothetical mass neutralizer does not qualify as an external agent—it's part of the object whose mass he is attempting to neutralize, and since I convinced him to use one of the lithium-ion battery packs from my laptop instead of its AC power converter, there is no wall cord to constrain the motion of his experiment. The amount of momentum and energy Luke's project has before he switches it on is exactly the same as it is afterward. An object's momentum is the product of its mass and velocity. Given that the Law of Conservation of Momentum says that an object's momentum—the product of its mass and velocity—is a constant in the absence of an external force, and since Luke has "magically" reduced his project's mass, the project's velocity *must* increase in order to conserve momentum.

While the Luke's experiment initially started at rest, recall that all motion is relative. So the experiment was motionless relative to the table on which it sat. The moment Luke throws the switch, however, his project immediately assumes a much higher velocity along a vector whose direction is defined by the rotation of Earth, the latitude of my kitchen table (approximately 39 degrees 43 minutes north) and the current time. Relative to the kitchen table, that speed would be approximately 823 miles per hour, slightly beyond the speed of sound. The project blasts through the top of the stove, punches a hole through the kitchen wall immediately below the window ledge, narrowly misses both dogs in the backyard before obliterating both my fence and the neighbors', takes the roof off the garage two houses down, and fells several trees and telephone poles in the next half mile—before the Earth's curvature removes any more breakable objects (such as the city of Beavercreek) from its flight path. F-16 pilots from nearby Wright-Patterson Air Force Base radio base that their planes, in full afterburner, lack the top speed to intercept...something. The switchboard at the Dayton Police Department is quickly flooded with UFO reports. Eventually the project returns to Earth's surface and, because of air friction, it crashes to the ground very hot, but at a mere couple hundred miles per hour.

While the example is deliberately tongue in cheek, the general concepts are still sound. However they may work, suspensors must obey momentum and energy conservation laws. Since no suspensor in any of Herbert's Dune novels has resulted in the type of high-speed projectiles of death like that resulting from Luke's hypothetical science fair project, we can probably rule out mass nullification as the operational method.

If a Holtzman field-based suspensor doesn't work by neutralizing an object's mass, perhaps it works by generating some sort of counterforce that opposes local gravity. This is an easy enough concept to understand: aircraft wings, helicopter rotors, rocket motors, hot air balloons, and blimps do this all the time in our world. Perhaps the low-drain phase of a Holtzman field does the same thing somehow.

As luck would have it, a solution like this exists that is within the scope of our current theoretical understanding. Dr. James F. Woodward is a physicist at California State University, Fullerton who has been publishing research in refereed scientific journals for over a decade on how Mach's Principle can be exploited to generate propulsive force without expelling either propellant.

The momentum argument can be described by a simple analogy to a car driving down the freeway. At rest, the car has no momentum, but in motion it has considerable momentum. But if we draw a dotted line around the entire planet, the TOTAL momentum is conserved. The planet has an unmeasurably small amount of extra momentum in the opposite direction to that of the car, and so momentum is conserved. In the case of a Mach Effect Thruster, the spacecraft being propelled by the thruster has increased momentum. But the rest of the matter in the universe has a correspondingly decreased momentum, so momentum is again conserved. We just need a bigger dotted line, encircling the entire universe.[1]

So the question becomes, can we generate propulsion, or momentum, for an object by drawing momentum from the rest of the Universe? The "Woodward Effect," as it has been called (Professor Woodward himself actually prefers the term "Mach Effect"), depends upon higher-order time-varying terms to the gravitational potential energy equation which normally have no effect or use in our everyday life. Restated, the equations defining how gravity behaves consist of several mathematical terms. In day-to-day applications, scientists and engineers use the first term only, and ignore all of what are called *higher-order terms*. For most applications, this is a very good approximation.

By generating a rapidly time-varying electromagnetic field, however,

---

[1] http://en.wikipedia.org/wiki/Woodward_effect.

it is theoretically possible to make these higher-order terms dominate the calculation for an object's gravitational interactions. The interesting thing about these higher-order terms is that they predict transient inertial mass fluctuations within the object. The more power we use in the electromagnetic field—and the higher the frequency with which we vary this field—the larger this mass fluctuation becomes.

Okay, we now have an object with a rapidly varying mass. How does this help us counteract local gravity and float above the planet's surface? Dr. Woodward and other researchers propose synchronizing the electromagnetic field with a piezoelectric transducer aimed along the direction in which you wish to move. Have the forward movement of the transducer occur at a moment when the electromagnetically induced mass fluctuation is making your object's mass lower. If the transducer and electromagnetic field are perfectly synchronized, this makes the backward movement of the transducer occur when the mass fluctuation is making your object more massive. Over one cycle, the entire system (electromagnetic field, mass, and transducer) will move forward much farther than it will move backward, just as if some force had acted on it to move it forward. If you have the device arranged correctly and are using the proper settings for field power and frequency, you can, in theory, generate enough force to oppose the local force of gravity.

The basics of the Woodward Effect have been horribly oversimplified here, but you should get the general idea. The more scientifically astute readers among you are probably thinking that a Woodward Device violates the conservation laws of momentum and energy to which I was so willing to sacrifice my kitchen and neighborhood earlier. It only seems like it is violating conservation because *you haven't defined your reference frame big enough*. Remember the earlier description of Mach's Principle: the inertia of a body (and thus its inertial mass) arises from its relationship to the totality of all other bodies in the universe. Your reference frame for calculating momentum and energy conservation cannot merely be Luke's device and the kitchen table. It has to be the *entire Universe*. When you define your reference frame that large, momentum and energy are conserved because the Woodward Device is transferring them from the rest of the Universe. Further, you need to look not at the mass change at a specific moment in time, but the average mass change over one cycle of a Woodward device. Since the net mass change is zero over the course of one cycle, the mass is always as it has been and I don't have to call the insurance claims adjuster, a handyman, or NORAD, after Luke switches on his new science fair project. The universe, its physical laws, and my neighbors can rest satisfied.

Generating a Woodward Effect with a Holtzman field would involve both generating a time-varying energy field of some sort and generating some sort of vibration effect parallel to the local gravitational field. Neither of these should take as much energy as interdicting directed energy fire or folding space-time—other applications of the Holtzman Effect stated in the Duniverse—so that should fit the "low-drain phase" portion of the suspensor definition in *Terminology of the Imperium*. The counterforce generated by the Woodward Effect will depend on the amount of mass for which you are compensating (Baron Harkonnen, please take note) and the amount of energy you are willing to consume. The bottom line is that the low-drain phase of a Holtzman field could very well be generating the Woodward Effect. While the concepts of time-varying mass and the Woodward Effect still rest largely in the realm of the theoretical, there have been several recent breakthroughs in the category of levitation, but all on a microscopic scale.

For years, until recently, astronomers have believed that the ultimate fate of the Universe would fall into one of two categories: heat death or recollapse. In the "heat death" scenario, the Universe would expand, and cool, forever. Ultimately there would not be enough energy to drive chemical, or biological, processes. In the recollapse scenario, the self-gravity of the Universe would ultimately pull all of the rapidly receding galaxies back together, in a scenario called the "Big Crunch." Recently, though, observational evidence has pointed to a third scenario: that the expansion of the Universe is literally accelerating, and that it may one day literally tear itself apart in what has been come to be known as the "Big Rip." The explanation for this acceleration is heretofore unexplained, but one of the likely suspects is something called the *Casimir Effect*.

The Casimir Effect is a result of what has been called *vacuum energy*—a form of energy that is pervasive throughout the Universe, but which science has yet been unable to harness (though references to tapping vacuum energy exist in science fiction, like the zero-point modules, or ZPMs, from *Stargate: SG-1*. It has been said that if all the vacuum energy within a common lightbulb could be harnessed, it could boil all of Earth's oceans.

The Casimir Effect is believed to result from the existence of vacuum energy. When two metallic plates are brought very close—approximately one hundredth millionth of a meter—vacuum energy between the plates creates a strong attractive force. This is called the Casimir Effect, or the Casimir Force. It has been postulated that the Casimir Force, or a related effect, can similarly be a repulsive force, and may be responsible for the accelerated expansion of the Universe. Scientists have also postulated that applications of the Casimir Effect can stabilize wormholes, thus allowing faster-than-light travel (again, see the FTL chapter, as well as *Stargate SG-1* and *Star Trek: Deep Space Nine*) (page 177). Recently, though, applications of the Casimir Force have allowed scientists to levitate microscopic objects. Is a macroscopic version in our near future? In concert with the Casimir Effect, scientists in the United Kingdom have also used metamaterials to generate repulsive forces. Metamaterials are substances whose properties are dictated less by their composition and more by the microscopic atomic structure. Metamaterials, coincidentally, have also shown promise toward practical applications of *invisibility*. So metamaterials show promise not only in real world suspensors, but also in cloaking devices.

Are suspensors potentially practical devices? Apparently so, believe it or not. We've examined several ways in which repulsive forces, hence suspensor technology from *Dune*, could be accomplished using the current state of scientific knowledge. Just remember, though, that if all this has given you an insatiable desire to experiment, please don't do it on your kitchen table. Your neighbors will thank you.

GES SEGER is the author of *The Once and Future War*. He is currently misusing both of his physics degrees working as a computer programmer at Wright-Patterson AFB in Ohio. Between that, writing, driving his wife and children to their activities, and his own competitive Irish dancing career, he has no time for anything in the way of meaningful hobbies.

References

F. Chen, G. L. Klimchitskaya, V. M. Mostepanenko, U. Mohideen, *Control of the Casimir force by the modification of dielectric properties with light.* Phys. Rev. B, v.76, N3, 035338-(1-15), 2007.

A. Einstein, letter to Ernst Mach, Zurich, 25 June 1923, in Misner, Charles; Thorne, Kip S.; and Wheeler, John Archibald (1973). *Gravitation.* San Francisco: W. H. Freeman.

Zeeya Merali, 6 August 2007, *Three ways to levitate a magic carpet* http://www.newscientist.com/article/dn12429-three-ways-to-levitate-a-magic-carpet.html

Mach's Principle. Wikipedia, the Free Encyclopedia. http://en.wikipedia.org/wiki/Mach%27s_principle

Woodward Effect. Wikipedia, the Free Encyclopedia.\ http://en.wikipedia.org/wiki/Woodward_effect

THE SHADE OF ULIET: MUSINGS ON THE ECOLOGY OF *DUNE*

David M. Lawrence

Arrakis. Dune. Desert planet. Arrakis has in abundance: sand, sand dunes, sand storms, sandworms, and sandtrout. It's easy to suppose that the ecology of a world like Arrakis would be fairly... parochial... in its scope. Scientist and journalist David M. Lawrence reveals that it is anything but.

The human question is not how many can possibly survive within the system, but what kind of existence is possible for those who do survive.

—Pardot Kynes

I F THE WAY BIOLOGISTS used to whisper about Frank Herbert's *Dune* is any indication, it was held in as high regard as the music of Thelonius Monk is held among jazz fanatics. They spoke of it in low, knowing whispers, nodding and waving their hands in a fashion that made one want to read the book in order to become as cool as they were.

Well, as cool as one biologist could look to another.

I first heard of *Dune* nearly thirty years ago, while an undergraduate at Louisiana State University in Shreveport. One of my mentors, a botany professor named Steve Lynch, repeatedly praised *Dune* as a work of ecological enlightenment. Reading the book, he said, was a life-changing experience.

Coming from Steve that said a lot. To a boy who grew up in northwestern Louisiana, Steve had the right credentials to know "cool." He was from California. He wore shorts and flip-flops to class. He played

217

guitar. He may or may not have inhaled. His golden Labrador retriever, Cooper, was smarter than many of my fellow students—and better company. And Steve was active in environmental causes.

He said I should read the book.

I filed away the suggestion, but beer, sex, and ZZ Top (not necessarily in that order) kept me too preoccupied to act for years. Nevertheless, I managed to catch every mention of it, no matter how far away, no matter how faint—like a radio telescope picking up transmissions of the value of π [pi] from somewhere in a galaxy billions of light years away. Still, I did not read the book. I paid great attention to the release of the movie, and did not see it—I wanted to read the book first.

But I did not, until recently.

The bulk of *Dune* is set on Arrakis, also known as Dune, a desert planet covered in massive sand seas with occasional rock-lined canyons, escarpments, and pavements. Hot, dry weather is broken up by vicious sand storms. Gigantic, hungry worms stalk the sands, preying on (or defending themselves against) inexperienced travelers that dare intrude upon the worms' domain. Arrakis is populated by one race of humans: ancient migrants, now long established and virtually native, that thrive in the wastes, expertly conserving the planet's most precious resource—water. These people are known as the Fremen.

Arrakis—the sole source of melange, or spice (the substance said to give its users prophetic powers), has important similarities to Earth. Like Earth, it is the third planet in its solar system. The Arrakean atmosphere is 75.4 percent nitrogen, 23 percent oxygen, 0.023 percent carbon dioxide, and trace gases; Earth's atmosphere is 78.1 percent nitrogen, 21 percent oxygen, 0.93 percent argon, 0.037 percent carbon dioxide, and trace gases. The chemistry of the Arrakean system is similar to that of Earth. The life is carbon-based. By inference, the availability of chemical building blocks for organic molecules is similar. In fact, as explained in the appendix to the first novel (*The Ecology of Dune*), plant growth is limited by the availability of trace minerals—fixed (organic) nitrogen and sulfur, in particular—much as plant growth on Earth is limited by availability of such minerals.

Nevertheless, the most limiting chemical constituent on Arrakis is water.

The portion of Arrakis most suitable for terraformation occured in a wide latitudinal belt ranging between 70 degrees north and 70 degrees south. On Earth, the most habitable portion occurs in a similar latitudinal band (roughly between 67 degrees north to 67 degrees south), with the boundaries determined largely by the location of the north and south polar fronts. As on Earth, the highest temperatures on Arrakis are found near the equator while the lowest are at the poles. The average temperatures on the Dune planet, however, are about the same as Earth, too. The average temperature at the Earth's surface is about 287 degrees K (what is referred to in *Dune* as absolute), or 14 degrees C (57 degrees F), whereas average temperatures in Earth's polar regions (beyond 66.5 degrees north or south latitude) range from near to well below freezing (less than 273 degrees K, 0 degrees C, or 32 degrees F).

Temperatures in the Arrakean polar regions are similar to that of Earth, ranging from near to well below freezing, thus permitting the existence of small icecaps that will potentially serve as important reservoirs of water. Between Dune's polar regions, however, temperatures range from 254 degrees to 332 degrees K, which is equivalent to a range of -19 degrees C to 59 degrees C or -2 degrees F to 138 degrees F. The tropical and middle latitudes of Arrakis have long growing seasons with temperatures in the 284 degrees to 302 degrees Kelvin range (11 degrees C to 29 degrees C or 51 degrees F to 84 degrees F), similar to that characteristic of the most productive terrestrial ecosystems of Earth.

Readers of science fiction and fantasy are expected to suspend disbelief. Disbelief is required when Herbert described the massive sandstorms that raged across the surface of Arrakis. One of the factors he said that contributed to the strength of the storms was the Coriolis effect. The effect, since it is often referred to as a "force," may seem a logical factor to consider in the strengthening of storm winds on the basis of the name. The Coriolis effect is not a true force, however. It adds nothing to wind speed. It is an apparent deflection in the path of a moving object, such as a storm system, when plotted on a rotating frame of reference. The object may actually move in a straight line, but it appears to follow a curved path as the frame of reference rotates out from under it.

For example, you and a friend are on opposite sides of a spinning merry-go-round. You are inspired to bean your friend with a tennis ball and throw it straight at him. You miss! The problem was not your aim. The problem was not the ball; it traveled in a straight line. The problem was that, as the merry-go-round spun, it carried your friend out of harm's way. If someone viewing the mayhem from a tower above plotted the path of the ball *on the merry-go-round*, the path would appear to be curved. On Earth, objects in the Northern Hemisphere appear to be deflected to the right. In the Southern Hemisphere, they appear to turn to the left. The spiral motion of winds around the eye of a hurricane is a product of the Coriolis effect. The speed of the winds around that eye is not. Wind speeds result primarily from the air pressure gradient as one moves from the eye outward. The steeper the pressure gradient (i.e., the greater the difference in pressure), the faster the wind speeds.

Herbert's description of the role of sandworms in the maintaining the oxygen concentration of the Arrakean atmosphere likewise requires substantial disbelief. The most widespread ecosystem on Arrakis is the desert. The sands of the desert are in part maintained and conditioned by the sandworms themselves. This is similar to the role of earthworms in your garden. The sandworms play a role in the generation of melange, which tiny sand plankton feed on, and sandworms in turn feed on the sand plankton. Herbert proposed a simple nutrient cycle, although it is more similar to that of the Earth's oceans than of any terrestrial ecosystem. The plankton are the producers, the sandworms are consumers (and somewhat omnivorous ones, too, as they feed on the sand plankton as well as on themselves). The sandworms, whether by their role in the production of melange, their own wastes, or the recycling of material from their own decomposition, produce nutrients for the plankton. With respect to nutrient cycling, so far, so good. The problem is in the cycling of oxygen. On Earth, the producers, typically photosynthetic plants and cyanobacteria, produce oxygen as a byproduct of the reaction that produces sugar from carbon dioxide and water. On Arrakis, the ultimate consumer, the sandworm, is responsible for oxygen production. An organism that can grow to as much as 400 meters long and travel through sand at speeds many fish can't achieve in water on Earth would seemingly consume a substantial amount of oxygen. That such an organism would become a net producer of oxygen seems implausible. Herbert wrote *Dune* more than a decade before the discovery of deep-sea ecosystems that are not based on photosynthesis, and therefore do not depend on the sun for survival. Had he known of such ecosystems, he might have devised a more credible explanation of the Arrakean oxygen cycle.

Unlike Earth, which is covered mostly by the oceans, Arrakis lacks much surface water. The desert planet is much more like Mars, with evidence of flowing water at some point in the distant past. While in the frigid temperatures of Mars, the evidence of surface water comes from the existence of fluvial landforms like gullies and canyons that on Earth are created by flowing water. In the case of torrid Arrakis, the salt pan discovered by Pardot Kynes, who was appointed by the Padishah Emperor Elrood IX as the first planetologist of Arrakis, is likewise evidence of water that had once been on the surface. Salt is not a product of volcanism. It is not a component typically found mixed with lava, for example. On Earth, salt is typically dissolved in water. Most is in the oceans, but some is in fresh water. When it is found in large deposits on land, such as in salt pans or salt domes, it is because the salt was left behind by water that at some point had evaporated away from the site.

What little surface water that existed by the time that Kynes arrived on Arrakis was found frozen in the tiny polar ice caps. The landscape is harsh, deadly for all but the hardiest, savviest inhabitants. Kynes sought to make the planet more livable for its human inhabitants. The desert conditions on Arrakis pose a seemingly insurmountable challenge: water is essential for life.

Despite the lack of water at the surface, Kynes discovered (before the events related in *Dune* take place) liquid water in the ground, and water vapor in the atmosphere above. He devised a plan to harness water and terraform the planet. Increased water at the surface will enable the establishment of a more and more diverse assemblage of species, ultimately leading to functioning ecosystems capable of sustaining themselves.

Because the Harkonnen overlords of Arrakis cared little about the livability of the planet (their priority is the production of melange), the elder Kynes launched the plan in secret, enlisting the Fremen, who longed for freedom from Harkonnen oppression. As he demonstrated himself to be a willing and efficient killer of the hated Harkonnen, the Fremen chose to follow Kynes. Their bond was cemented through his marriage to a Fremen woman and fathering Liet-Kynes. The younger Kynes fulfilled all the rites of passage expected of Fremen males and was fully accepted by Fremen society. In fact, both the elder and younger Kynes became semi-mystical leaders of the Fremen.

Herbert's keen grasp of ecology is evident in his description of the elder Kynes's plans of how to terraform the Arrakean landscape. The transformation began with the construction of wind traps to intercept water vapor from the air and underground basins to store the collected water. With sand surface temperatures ranging from 344 degrees to 350 degrees Kelvin (71 degrees C to 77 degrees C or 159 degrees F to 170 degrees F) and with hot air temperatures (46 to 52 degrees C or 114 to 125 degrees F) above, one can expect ready loss of surface water to the atmosphere through evaporation unless there is some way to shield the water supply from the heat. Then the greening of Arrakis began. Water-efficient species that also served as excellent ground cover were planted first. As the dunes were stabilized with the plant cover protecting the dune surface from wind erosion, weedy species were planted next, followed by low perennials, and then larger perennials: cactuses, shrubs and trees tolerant of arid conditions. The ultimate goal was to cultivate food and medicinal species.

The elder Kynes's plan had a logical progression. Each cohort of plants will help stabilize the sand surface, capture more water from the atmosphere and retain it in the biomass and ground, and transform the mostly mineral sand into true soil by enriching it with organic matter and nutrients. The success of one cohort will make the changed environment suitable for subsequent plantings. Kynes recognized that plants cannot succeed on their own. They needed animals, both invertebrates and vertebrates, to help work the soil, contribute to nutrient cycling, pollinate flowers, and graze to prevent fecund and fast-growing species from the others.

The elder Kynes recognized one other important fact: his plan would fail unless it was sustained over the centuries. As the Fremen were his only hope for carrying out the plan, he had to educate them in ecological matters. His educational program skillfully exploited their religious beliefs, binding his ecological principles with their traditional ethical and moral principles to create an environmentally enlightened society. He groomed his son as his successor. A wise move, as he was killed prematurely in a cave-in at the Plaster Basin. His son, appointed his successor as planetologist of Arrakis by Padishah Emperor Shaddam IV, was continuing his terraforming program at the time Duke Leto Atreides assumed the fiefdom of the planet at the behest of the Padishah Emperor.

> The effect of Arrakis on the mind of the newcomer usually is that of overpowering barren land. The stranger might think nothing could live or grow in the open here, that this was the true wasteland that had never been fertile and never would be.
>
> —From *The Ecology of Dune*

The elder Kynes's plan may seem far-fetched—terraforming a planet—but humans have terraformed landscapes for millennia. Unless one includes an unproven (though not far-fetched) hypothesis that pre-agricultural humans opened up the landscape to make it more suitable for hunting by harnessing fire, the first major transformation of the landscape was the transformation of steppes and forests into farmland at the dawn of civilization. Irrigation and cultivation are hallmarks of human settlement in arid regions. The kind of transformation Herbert envisions in *Dune* is far grander and much more fundamental than anything attempted on Earth.

Or is it?

In March of 2007 I joined a biological expedition to Ascension Island, the peak of a volcano rising from the mid-Atlantic Ridge in the South Atlantic Ocean. It is small, only about ninety square kilometers (thirty-six square miles), with its highest point about 860 meters (2,817 feet) above sea level. It is not a classic desert island, i.e., not the kind of place envisioned by people "informed" by the writings of Robert Louis Stevenson, such as *Treasure Island*, or movies like *Pirates of the Caribbean*. It is not a choice destination for people whose ideas of roughing it consist of sun, sand, and bizarre alcoholic concoctions topped by miniature umbrellas. Given that Ascension is home to both British and American military installations, thus populated by mili-

THE SCIENCE OF DUNE

tary personnel and contract workers, and given that there are few recreational opportunities (there is only so far one can drive), I am sure
one can find plenty of bizarre alcoholic concoctions. There is no short-
age of sun and sand. There also is, unlike Arrakis, plenty of water.
The nearest land is Napoleon's final home of St. Helena and is about
1,200 kilometers (750 miles) away. But the ocean water is of no use for
drinking (without desalinization) or irrigation. Ascension is located
near the equator, thus hot with few breaks in the weather. There are
none of what most people outside the tropics call seasons.

Charles Robert Darwin visited Ascension in 1836, near the end of
the expedition that provided the seed for what became his theory of
evolution by natural selection. He aptly describes Ascension in his
memoir, *Voyage of the Beagle*:

> On the 19th of July we reached Ascension. Those who have beheld a
> volcanic island, situated under an arid climate, will at once be able to
> picture to themselves the appearance of Ascension. They will imagine
> smooth conical hills of a bright red colour, with their summits generally
> truncated, rising separately out of a level surface of black rugged lava. A
> principal mound in the centre of the island, seems the father of the less-
> er cones. It is called Green Hill: its name being taken from the faintest
> tinge of that colour, which at this time of the year is barely perceptible
> from the anchorage. To complete the desolate scene, the black rocks on
> the coast are lashed by a wild and turbulent sea....
>
> Near this coast nothing grows; further inland, an occasional green
> castor-oil plant, and a few grasshoppers, true friends of the desert, may
> be met with. Some grass is scattered over the surface of the central el-
> evated region....

Joseph Dalton Hooker, a friend of Darwin's who was one of the leading
botanists, as well as one of more eloquent advocates for the theory of
evolution, also visited Ascension in 1843 while accompanying James
Clark Ross's expedition to Antarctica. In an 1866 article entitled "Insu-
lar Floras," Hooker wrote that the island "consists of a scorched mass
of volcanic matter, in part resembling bottle glass, and in part coke
and cinders. A small green peak, 2880 feet above the sea, monopo-
lises nearly all the vegetation, which consists of purslane, a grass, and
a euphorbia in the lower parts of the island, whilst the green peak is

clothed with a carpet of ferns and here and there a shrub allied to but different from any St. Helena one." Probably only about three dozen plant species could be found on Ascension at the time of its discovery. With the exception of seabirds, a flightless rail, and sea turtles, the largest animals native to Ascension were land crabs and insects.

Further confirmation of Ascension's desolation is in the fact that the Portuguese, who discovered the island in 1501, uncharacteristically deemed it unworthy of colonization. The British finally settled the island in 1815 for strategic reasons alone. After Napoleon's exile to St. Helena, they did not want the French to use the island as a base from which to liberate the conquered emperor. Once there, the British decided to keep a garrison on the island, but struggled to make it habitable. Hooker, after the Ross expedition, devised a plan to achieve that goal, thus becoming the Pardot Kynes of Ascension history.

In 1847 Hooker prepared a report for the British Admiralty outlining a program of terraforming (long before the term was coined) Ascension. As with the elder Kynes's plan for Arrakis, it was multifaceted. Hooker suggested that rainfall on the island could be increased by planting large trees on the upper slopes of Green Mountain (Darwin's Green Hill). Steep ravines should be planted with shrubs, low trees, and cactuses. He felt these woody species would promote soil formation and reduce evaporation of moisture from the soil surface. Valleys at lower elevations should be planted with appropriate species, such as acacia, casuarina, eucalyptus, and other plants adapted to arid conditions. Finally, tropical and European crops should be planted in gardens high on Green Mountain. Hooker recommended some species that could be introduced and outlined where those species could be obtained. (Many were supplied by the Royal Botanic Gardens at Kew, to which Hooker succeeded his father, William Jackson Hooker, as director.) Plants had been imported haphazardly prior to Hooker's report. Even less planning was devoted to animal introductions. Some imports, such as rats and mice, were desired by no one. Hooker's report provided a sense of purpose and organization to the effort to transform Ascension.

The Royal Navy and Marines followed the recommendations of Hooker, employing a succession of gardeners to carry out the planting program. The first two gardeners did their best, but life on the island proved too much and they (or in one case, a wife) helped engineer pretexts to get them returned home to England. While their efforts to make the bare slopes of the cinder cones "productive" as the Royal Navy requested failed, Green Mountain was slowly transformed. The transformation flourished under the guidance of the third gardener, a civilian named John Bell. Under his regime agricultural production from the gardens on Green Mountain increased by 50 percent to 90,000 pounds. He planted tens of thousands of trees and shrubs.

Plants were brought in from all parts of the world: Buddleia, chasteberry, greasy grass, and periwinkle from Africa; acacia, agave, aloe, and prickly pear cactus from the Americas; bamboo, banyan, and ginger from Asia; blackberry and raspberry from Europe; casuarina and eucalyptus from Australia; and screw pine and coffee wood with worldwide or nearly worldwide distributions. Animals introduced to Ascension included cats, donkeys, goats, sheep, and a number of birds, including, for example, the Indian mynah.

By the time I arrived at Ascension, the summit of Green Mountain had been transformed from a rocky prominence with a faint green cap of ferns to a diverse, artificial—yet functional—forest. A thick stand of bamboo blankets the highest elevations. Surrounding the bamboo, especially on the northern slopes of the peak, is a forest consisting of tall specimens of eucalyptus, mulberry, palm, screw pine, yew, and others. The cloud forest at the summit grades into expanses of grass and scrub forest at lower elevations. Greasy grass and guava are common on drier slopes. Bermuda cedar and casuarina create something of a forest in more humid locations. A spectacular grove of Norfolk Island pine occurs on the southeast slope. At lower elevations, wide expanses of a stark, charred volcanic landscape prevail. Indigenous plant species can be found in gullies fed by occasional heavy rains, but a number of introduced species have gained a toehold there as well, including castor oil, horse tamarind, waltheria, and the yellow thistle. Prior to the establishment of the British garrison, the lower elevations outside the gullies were bare except for brief periods after heavy rains when pappusgrass flourished. Now, mesquite (called Mexican thorn on Ascension) and prickly pear, staples of American deserts, have spread and are helping transform bare volcanic rock in to true soil.

While the introduction of many plants to Ascension followed something of a plan—Hooker's—the introduction of animals to the island had a far more haphazard history. Rats (both black and Norway) and house mice were accidentally introduced from ships (it should be no surprise that no one wanted them, either at sea or on shore). Goats were probably brought to the island shortly after its discovery as a source of food for meat-starved sailors (this was common practice in the age of sail). With the British attempt to launch agriculture in the nineteenth century came agricultural pests. Birds (starlings, thrushes, rooks, jackdaws, common waxbills, yellow canaries, and the Indian mynah) were introduced to help control insects. Hedgehogs were also brought in to control agricultural pests. Cats were released in the hopes that they would help control rats, as were barn owls. Rabbits were imported as game. Cattle and sheep arrived with the British garrison in its attempt to establish agriculture. Donkeys were brought in as beasts of burden.

The introductions have a mixed record. Some of Hooker's vision has been borne out—especially by the transformation of the summit of Green Mountain into cloud forest. Whether or not the forest has led to increased rainfall is arguable, but the vegetation serves moisture captured from the southeast trade winds, keeping the summit enshrouded in clouds. (Ironically, the development of the forest may have contributed to the drying of springs on the slopes of Green Mountain, as water that used to bubble out of the ground is now pumped into the atmosphere through transpiration, loss of water through stomates, or openings, in the leaves of the plants.) Hooker's plan has likewise worked for the lower elevations as species like mesquite and prickly pear build soil on what was once a barren, ash- and rock-covered surface. These plants are often candidates for control by Ascension conservation authorities. The downside of this transformation has been the extinction of some endemic species: plants that were known only to Ascension. Other endemic species have been reduced in number, or their ranges on the island have gotten smaller and smaller as a result of competition with introduced plants.

Introduced plants had an easier time of it on Ascension than introduced animals. A number of the bird species introduced during the nineteenth century failed to gain a toehold on the island. Only the canary, francolin, mynah, and waxbill had successfully established themselves. Another import, the house sparrow, has gained a toehold on the island in the twentieth century, primarily in developed portions of the island.

The larger animals such as goats, sheep, cattle, and donkeys have significantly influenced the plant community, helping to expand the range of introduced plant species on Ascension at the expense of natives. While goats were eliminated by the middle of the twentieth century, and cattle were nowhere to be seen during my expedition in 2007 (although one is reported to survive on Green Mountain), donkeys and sheep are still quite common. In recent decades both have contributed to the rapid spread of mesquite, a recent introduction to the island. Introduced animals, primarily rats and the cats brought in to control them, have often had a devastating effect on native animals, particularly seabirds and sea turtles that breed there, by preying on the nests, eating eggs and hatchlings. The flightless Ascension rail was probably rendered extinct by predation from rats before the British garrisoned the island in 1815. The combined pressure by cats and rats has driven many seabirds from their roosting areas on the island, but the recent elimination of feral cats has allowed some of those species to return.

As with Kynes's plans for Arrakis, the reason for terraforming Ascension was to make the island more livable. A primary goal was to increase the available water supply, and one way to do that is to capture moisture available from the atmosphere. As on Arrakis, the amount of water available in the air on Ascension increases with elevation, especially on the southern and eastern slopes of Green Mountain, which are exposed to the southeast trade winds. Thus, the British garrison constructed its own version of a water collector. They paved the head of Breakneck Valley, a long, southeast-facing valley that runs from the summit of Green Mountain to its base, and installed a collection basin and piping at the lower portion of the pavement to gather and transport water to the island's settlements.

As the history of Ascension Island bears out, terraforming even a relatively small landscape is difficult. It is difficult to consider the magnitude of the task the elder Kynes set for himself and the Fremen—to embark on a three- to five-century-long project to terraform an entire planet—without a shudder. The large area to be transformed, the harsh initial conditions, and the dearth of raw materials make success unlikely. The consequences of failure were high: too much water could poison the sandworms, thus destroying melange production as well as a major source of oxygen for the planet's inhabitants. But Kynes embarked on the project with a faith that justified his status as a holy man among the Fremen.

As with Hooker's recommendations for Ascension, the elder Kynes's plans for Arrakis drew upon a diverse pool of plant and animal resources for transformation of Ascension. The process begins with the planting of poverty grasses, which in their native North America grow on sandy, impoverished soils (such as coastal dunes). Through their extensive root systems these plants bind the sand grains and make the soils less susceptible to erosion by wind or water. Once the poverty grasses have minimized the wind erosion, hence migration, of the Arrakean sand dunes, sword grasses (a term that refers to a widespread but taxonomically diverse array of species on Earth) will be used to further stabilize the dunes.

Once the physical stabilization of the dunes is achieved, the next task is the transformation of the mostly mineral sand to more organically enriched (thus more fertile) soil. Kynes started that with an array of ephemeral (weedy) species, such as pigweeds, a collective term that includes both *Amaranthus* and *Chenopodium*. As weeds, they can grow on infertile sites. As such species germinate, grow, reproduce, and die, they chemically alter the soil around them via reactions in the root zone as well as by adding organic matter through decomposition.

The next phalanx of plants to be planted by Kynes included scotch broom, low lupine, dwarf tamarisk, and shore pine as well as the classic desert species: barrel cactuses, candelilla, and saguaro. Scotch broom, which is native to Europe and Africa, and low lupine, native to North America, are legumes. As such, they, with the help of nitrogen-fixing bacteria in their roots, naturally fertilize the soil with organic nitrogen that can be taken up by other species. Dwarf tamarisk, a native of arid lands in Africa and Eurasia, and shore pine, native to North America, are both tree species that do well on dry, nutrient-poor soils. Cactuses are native to the Americas, commonly found in desert regions. Both barrel cactuses and saguaro are succulents, e.g., they store water in their stems. (Saguaro is the giant of the America deserts, growing to more than 13 meters [forty-two feet] tall.) Cactuses have green stems and lack leaves. This allows them to make their own food (sugars) via transpiration while minimizing the loss of water via transpiration. Candelilla, while native to North America, is a member of the spurge family, whose species often play a similar role in Eurasian and African deserts as the cactus family does in the Americas. Succulent spurges even look like cactuses, with green stems and reduced or no leaves to allow photosynthesis while limiting water loss.

In suitable areas, Kynes recommended the planting of other species, all of which can be found on Earth in arid and semi-arid environments. These included onion grasses, which have a widespread distribution; Gobi feather grass, from the steppes of Eurasia and Africa; camel sage, a perennial herb found in Eurasia and Africa which is known for its sand-stabilizing abilities; evening primrose, found in dry and disturbed environments of North America; sand verbena, a creeping plant from the deserts and other sandy environments of the Americas; wild alfalfa, a legume native to grasslands and woodlands of North America; the burrow and creosote bushes, shrubs in desert North America; and incense bush, a name that refers to a number of unrelated shrub in both of the Americas.

Kynes knew that a functional ecosystem consisted of more than just plants, so he introduced animals as well. Just as the plants, the candidates for animal introductions were selected with a specific purpose in mind. Herbivores, for example, keep the plant population in check in a functioning ecosystem. Without them, some plant species would overgrow the others. Carnivores keep the herbivore population under control. Without predators, the herbivores would increase in numbers and overgraze the vegetation. Kynes first selected burrowing animals to work the soil: the kangaroo mouse, an herbivore, the kit fox, a carnivore, both natives of North America, and the sand terrapin, an herbivore, possibly the desert terrapin of the southwestern United States. Another herbivore, the desert hare, could either be a Eurasian species or else a species also known as the black-tailed jackrabbit of North America. He chose other vertebrate predators to help manage the herbivore population: desert, dwarf, and eagle owls, and desert hawks (each with a number of possible Earth analogs). Kynes did not ignore the invertebrate world, adding familiar residents of arid environments:

the biting wasp, centipede, scorpion, trapdoor spider, and wormfly, and adding desert bats to control the invertebrates (although it is unlikely that bats would normally prey on most of the invertebrates mentioned).

Once the transformed Arrakean ecosystem proved it could function, Kynes planned to introduce food crops such as date palms and melons, coffee, fiber crops such as cotton, and medicinal plants.

In the Dune universe, the elder Kynes's plan succeeded so well that 200 years after Paul Atreides, son and heir of Duke Leto Atreides, defeated the Imperial and Harkonnen forces that conspired to kill his father (and who also murdered Liet-Kynes by abandoning him in the desert), the waters that once covered the surface of Arrakis returned and drove the sandworms and sandtrout (a larval form of sandworm) to extinction.

Kynes's plan was carefully thought out, tested, revised, and tested again. Terraforming does not have to proceed in such a logical fashion to work, however. Hooker had a plan for Ascension Island, but much of the terraforming work was done by accident, without thinking. Mistakes were made, as in the devastating effect that cats brought in to control rats had on the native bird life. But parts of the effort worked better than expected.

Under normal conditions, ecosystems are assembled over time in a process called succession. In the case of relatively bare land, succession typically begins with a few fast-growing species that modify the environment, such as adding nutrients to the soil and making conditions favorable to other species. The later arrivals continue to modify the environment, and new species colonize the system until a point where the amount of living matter and often the diversity of the system are maximized. Species that fit in with others present become successfully established. Those that don't cannot gain a toehold and fail. In many ecosystems, this process takes centuries. (In truth, it never stops; change is a constant in living systems.)

While there are significant differences in scale between the task undertaken by Kynes to terraform an entire planet and that of the British in terraforming Ascension, I think the two efforts can be realistically compared. If anything, Ascension, with its arid environment, can be considered a microcosm, miniature, of Arrakis. In *Dune*, Kynes's plan was supposed to take several centuries to unfold, but it lacked much of the natural trial and error of succession. On Ascension, the process has taken less than a century, but with a significant amount of successional experimentation. That experimentation continues, demonstrated by the march of mesquite and prickly pear across the bare lowlands of Ascension. While likely driving some of the island's endemic species to extinction, the development of the artificial forest has made Ascension a place where people who have no right of residency under current British policy continually renew employment contracts that keep them on the island for decades. An island described by Darwin in 1839 as "not smiling with beauty, but staring with naked hideousness," was transformed by 1920s into a place, as marine biologist Alistair Hardy wrote, where the "colours are fantastic," and "sheep grazed on the slopes of grass in between patches of almost dense jungle." But both Ascension and *Dune* illustrate the dangers of creating paradise. Some of Ascension's endemic plant and animal species have been driven to extinction as a result of the changes humans have made, such as the Dune saga, where the sandworms—arguably the most important species on the planet—are driven to extinction.

Now I know why people like Steve Lynch whispered so enthusiastically about *Dune*. Herbert's vision, as epic and fantastic as it is, is grounded in ecological reality. While reading a futuristic tale of imperial intrigue, religious mysticism, and desert adventure, we can learn something about taking care of the place we call home.

DAVID M. LAWRENCE has never decided what he will do when (if) he grows up. He is a scientist who teaches geography, meteorology, oceanography, and (sometimes) biology at the college level. He is a journalist who covers everything from high school sports to international research in science and medicine. He is a SCUBA diver looking for a way to make a living on the water. When not consumed with those activities, he looks at his guitars and wonders if he's too old to become a rock god. (It would help if he could actually play.) He lives in Mechanicsville, VA, with his wife, two children, and a menagerie of creatures with legs, scales, and fins.